Modelling and Control Design of Vibration Reduction Systems

Tomasz Krzyzynski · Igor Maciejewski
Lutz Meyer · Henning Meyer

Modelling and Control Design of Vibration Reduction Systems

Methods and Procedures of Selecting Vibro-isolation Properties

Springer

Tomasz Krzyzynski
Faculty of Technology and Education
Koszalin University of Technology
Koszalin, Poland

Igor Maciejewski
Faculty of Technology and Education
Koszalin University of Technology
Koszalin, Poland

Lutz Meyer
Test Laboratory
Isringhausen GMBH and CO. KG
Lemgo, Nordrhein-Westfalen, Germany

Henning Meyer
Test Laboratory
Isringhausen GMBH and CO. KG
Lemgo, Nordrhein-Westfalen, Germany

ISBN 978-3-030-03046-9 ISBN 978-3-030-03047-6 (eBook)
https://doi.org/10.1007/978-3-030-03047-6

Library of Congress Control Number: 2018960740

This Springer imprint is published by the registered company Springer Nature Switzerland AG
The registered company address is: Gewerbestrasse 11, 6330 Cham, Switzerland

Preface

Scope

The aim of the book is to develop and verify appropriate methods and procedures for selecting the vibro-isolation properties of vibration reduction systems which is enabled to evaluate the visco-elastic characteristics of passive systems and support the control system design of semi-active and active systems. The generalised methods and procedures should allow to select the system vibro-isolation properties for different classes of the oscillatory systems in terms of the chosen optimisation criteria (usually conflicted criteria). The elaborated methods and procedures enable to protect the working machine operators from the adverse effects of vibration exposure.

Content

At first, spectral classes of the excitation signals are determined that are representative for different vibrating systems. This is a basis of shaping the vibro-isolation properties of vibration reduction systems. Then, the generalised mathematical models of vibration isolation systems are developed. These models enable to shape non-linear characteristics of the visco-elastic elements included in vibration isolator. Such a mathematical description of the vibration reduction systems determines the ability to use the developed models in a wide range of applications. Successively, the reliable and efficient optimisation criteria are determined and these criteria are frequency weighted in order to indicate the resonances of human body parts and organs.

Secondly, a method of shaping the vibro-isolation properties of passive vibration reduction systems is elaborated and such a method allows to determine non-linear characteristics of the basic elements applied in the system. This is achieved using the optimisation procedures that are able to find compromise solutions

(Pareto-optimal) with respect to conflicted vibro-isolation criteria. The effectiveness of proposed method is verified experimentally using the exemplary, passive suspension system that is excited using the signals of various spectral classes.

Finally, a method of shaping the vibro-isolation properties of semi-active and active systems is elaborated and such a method facilitates the control design of technically advanced vibration reduction systems. Using the proposed optimisation procedures, it is possible to adjust the vibro-isolation properties by changing the controller settings. The designed control system using the proposed method is investigated experimentally by means of the exemplary, semi-active and active suspension systems that are loaded of different masses.

Acknowledgements

The following work is a part of the research project 'Methods and procedures of selecting vibro-isolation properties of vibration reduction systems' founded by the National Science Center of Poland under the contract No. UMO-2013/11/B/ST8/03881. We would like to thank ISRINGHAUSEN GmbH & Co. KG for the assistance in the experimental research.

Koszalin, Poland
September 2018

Tomasz Krzyzynski
Igor Maciejewski
Lutz Meyer
Henning Meyer

Contents

1 Introduction 1
1.1 State of the Art 1
1.2 Research Methodology 5
References 7

2 Simulated Input Vibration 9
2.1 Nomenclature 9
2.2 Introduction 10
2.3 Random Signal Generator 11
2.4 Signal Processing Technique 12
2.5 Spectral Estimation Method 15
2.6 Vertical Vibration 17
2.7 Horizontal Vibration 18
2.8 Conclusions 27
References 27

3 Modelling of the Vibration Reduction System 29
3.1 Nomenclature 29
3.2 Introduction 30
3.3 Generalised Model of the System 30
3.4 Non-linear Models of the Typical Suspension System Components 33
3.4.1 Mechanical Spring Model 33
3.4.2 End-Stop Buffer Model 36
3.4.3 Hydraulic Damper Model 38
3.4.4 Friction Model 40
3.4.5 MR Damper Model 41
3.4.6 Pneumatic Muscle Model 43
3.5 Example: Modelling and Simulation of a Horizontal Seat Suspension 46

3.5.1 Model of the Passive Suspension 46
3.5.2 Evaluation of the Model Parameters 47
3.5.3 Model Verification 49
3.6 Apparent Mass Model of the Seated Human Body 52
3.7 Example: Models of the Apparent Mass for Seated Human Body 56
3.7.1 Apparent Mass Model Under Vertical Vibration 56
3.7.2 Apparent Mass Model Under Horizontal Vibration 58
3.8 Conclusions 61
References 62

4 Vibro-isolation Criteria 65
4.1 Nomenclature 65
4.2 Introduction 66
4.3 Evaluation of the Human Exposure to Whole-Body Vibration 67
4.4 Evaluation of the System Vibro-isolation Properties 72
4.5 Vibro-isolation Properties of the Horizontal Seat Suspension 75
4.6 Laboratory Evaluation of the Operator Seat Vibration 78
4.7 Conclusions 83
References 83

5 Multi-criteria Optimisation of the Vibro-isolation Properties 85
5.1 Nomenclature 85
5.2 Introduction 85
5.3 Problem Formulation 86
5.4 Global Sensitivity Analysis 88
5.4.1 Sample Matrix 88
5.4.2 Variance-Based Method 89
5.4.3 Sensitivity Indices 90
5.5 Example: Sensitivity Analysis of the Horizontal Seat Suspension 91
5.6 Optimisation Procedure 93
5.6.1 Decision Variables 93
5.6.2 Random Starting Points 93
5.6.3 Minimisation of Particular Optimisation Criteria 94
5.6.4 Suspension Travel Range 94
5.6.5 Minimising Both of the Conflicted Vibro-isolation Criteria 95
5.7 Example: Multi-criteria Optimisation of the Horizontal Seat Suspension 96
5.8 Experimental Research of the Optimised Horizontal Suspension with Seated Human Body 100
5.9 Conclusions 103
References 103

6 Computational Method of Selecting Vibro-isolation Properties 105
6.1 Nomenclature 105
6.2 Introduction 106
6.3 Formulation of the Overall Method 106
6.4 Random Input Signals 107
6.5 Simulation Model of the Passive System 109
6.6 Vibro-isolation Properties 110
6.7 Multi-criteria Optimisation of the Vibro-isolation Properties 113
6.8 Selection of the Visco-elastic Characteristics of Horizontal Seat Suspension for Chosen Excitation Signals 117
6.8.1 Random Vibration Inputs Affecting the Operators of Different Machineries at Work 117
6.8.2 Sensitivity Analysis of the Passive System for Various Visco-elastic Characteristics 117
6.8.3 Multi-criteria Optimisation of the Vibro-isolation Properties 122
6.8.4 Experimental Verification of the System 125
6.9 Conclusions 128
References 129

7 Control System Design 131
7.1 Nomenclature 131
7.2 Introduction 131
7.3 System Structure 132
7.4 Primary Controller 133
7.5 Secondary Controller 135
7.6 Example: Control System of a Horizontal Seat Suspension with the MR Damper 137
7.6.1 Model of the Semi-active Suspension 137
7.6.2 Control System Synthesis of the Semi-active Suspension 138
7.6.3 Semi-active System Verification 140
7.7 Example: Control System of a Horizontal Seat Suspension with the Pneumatic Muscles 142
7.7.1 Model of the Active Suspension 142
7.7.2 Control System Synthesis of the Active Suspension 144
7.7.3 Active System Verification 149
7.8 Conclusions 153
References 153

8 A Method of Shaping the Vibro-isolation Properties of Semi-active and Active Systems 155
8.1 Nomenclature 155
8.2 Introduction 155

8.3 Formulation of the Overall Method 156
8.4 Simulation Model of the Semi-active and Active Systems 157
8.5 Control System 158
8.6 Example: Control System Design of the Horizontal Suspension with Seated Human Body 161
8.6.1 Physical and Mathematical Model of the System 161
8.6.2 Optimisation of the Controller Settings 164
8.6.3 Laboratory Investigation of the Semi-active and Active Seat Vibration Control 169
8.7 Selection of the Vibro-isolation Properties of Semi-active and Active Systems for Chosen Excitation Signals 170
8.8 Conclusions 174
References 179

9 Conclusions 181
9.1 Summary of the Most Important Results 181
9.2 Final Conclusions 182
9.2.1 Conclusions Regarding the Developed Models of Vibration Reduction Systems 182
9.2.2 Conclusions Regarding the Developed Methodology of Shaping the Vibration Isolation Properties 183
9.3 Directions for Further Research 184

Chapter 1
Introduction

1.1 State of the Art

There are two basic sources of mechanical vibrations that can disturb a proper functioning of the machines. The first one relates to the systems that during their operation generate vibrations (Fig. 1.1a), e.g. engine vibrations in the working machine. The second class concerns the systems that vibrate due to external reactions (Fig. 1.1b), e.g. cab vibrations of the machine which is moving over uneven ground [1, 2]. Many sources of the vibrations can cause periodic or random reactions of the machine elements. In such situation, resonant states can be obtained and they may cause movement disturbances of individual machine elements. The resonant vibrations have an adverse effect on machine functioning, and this can lead to their failure. In addition, the vibrations have a negative influence on the health condition of working machine operators [3].

In most cases, the vibrations are harmful processes, having a detrimental impact on the human operator of a machine. The danger arising from the exposure of a human body to vibrations increases in the case of high-intensity vibrations with a prolonged duration period. Figure 1.2 gives the root mean square (RMS) acceleration values of the vibrations to which the operators of the most popular heavy machinery are exposed during their work [4]. The data presented are only to visualise the ranges of accelerations of the vibrations that induce operators' movements.

Earth-moving machines and most other forms of engineering vehicle expose their operators to whole-body vibration. According to the ISO-2631 standard [5], the criterion employed to evaluate the intensity of vibration transmitted to a human is based on the length of time (in hours) over which an exposure to vibration results in no health risks (NH), potential health risks (PR) or likely health risks (HR). For better presentation, the curves defining these three levels of the vibration exposure are shown in Fig. 1.3.

There are several methods to minimise vibrations which have a negative influence on working machines operators during their work. The first group of methods

T. Krzyzynski et al., *Modelling and Control Design of Vibration Reduction Systems*, https://doi.org/10.1007/978-3-030-03047-6_1

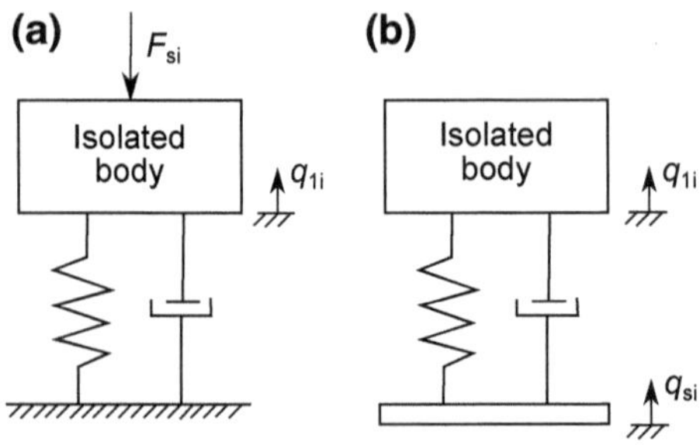

Fig. 1.1 Operating equipment generating a disturbance force F_{si} (**a**) and sensitive equipment supported by a vibrating structure q_{si} (**b**)

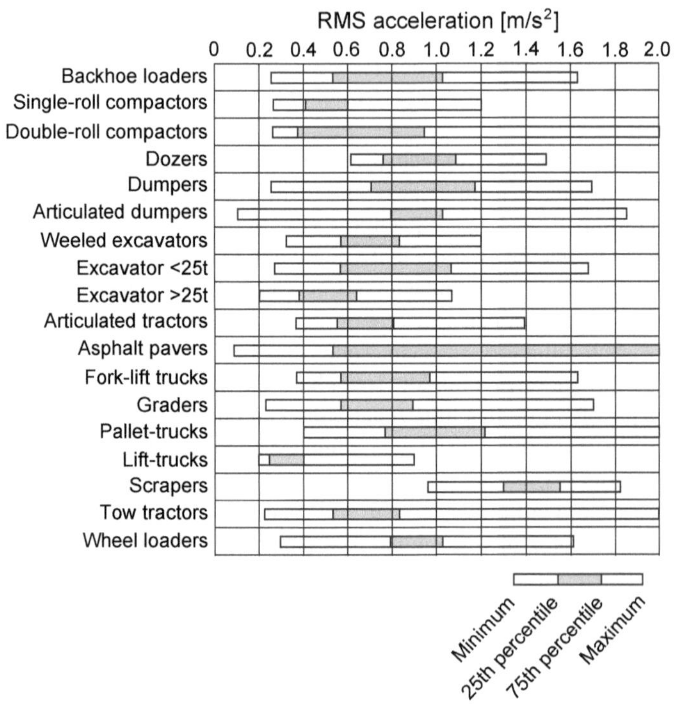

Fig. 1.2 Root mean square (RMS) acceleration of vibration obtained from more than 20 measurements carried out for each of the machine types [4]

comprises the passive techniques, and these techniques can be introduced in the following way [6]:

- prevention of vibration,
- structural modifications,
- parametric modifications,
- damping of vibration.

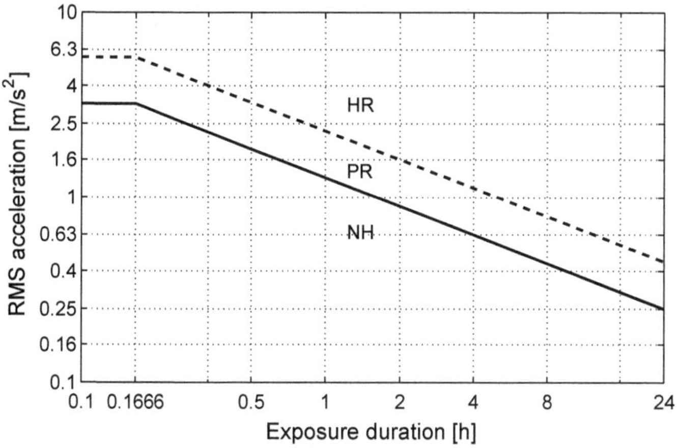

Fig. 1.3 Maximum vibration exposure time (hours) that corresponds to no health risk condition (NH), potential health risk (PR) and health risk is likely (HR) in accordance with the ISO-2631 standard [5]

In the recent literature, [3, 7, 8] only simplified, linear models of the vibration reduction systems are discussed. However, it should be noted that many vibration isolators demonstrate non-linear stiffness (e.g. pneumatic spring) and damping characteristics (e.g. hydraulic shock-absorber). The stiffness of suspension system is non-linear in the relative displacement domain, and its characteristic has a dissimilar shape for different loading masses (Fig. 1.4a). The damping of suspension system is also non-linear in the relative velocity domain, and its characteristic can be changed by the machine operator using the control lever (Fig. 1.4b). These considerations lead to the conclusion that linear mathematical description of many modelled vibration isolators is possible only with some approximation around a system operating point.

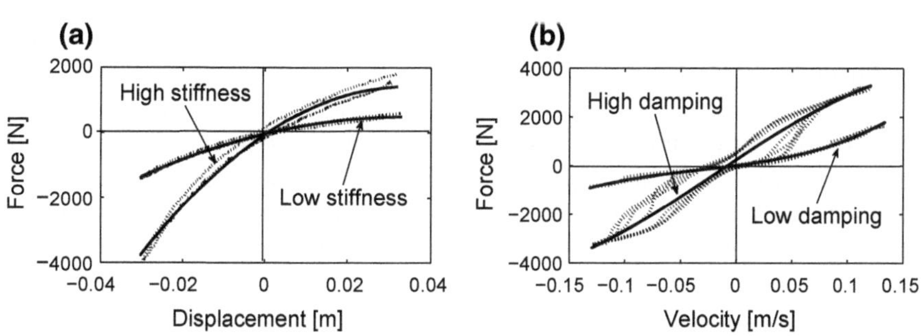

Fig. 1.4 Exemplary stiffness characteristic of the pneumatic spring: measurement (⋯⋯), approximation (—) and damping characteristics of the hydraulic shock-absorber: measurement (⋯⋯), approximation (—) [9]

There are no effective methods for shaping the vibro-isolation properties of the vibration reduction systems having non-linear visco-elastic characteristics which should be chosen for specific spectral classes of the excitation signals. A vibration isolator should be designed in such a way that efficiently reduces the vibrations in whole frequency range of the excitation signal (most often random excitation). Unfortunately, the optimisation of vibration isolators is very rarely used during a design process because the ongoing works have difficulties connected with the computational problems [10]. The low efficiency of the optimisation procedure is related to a significant number of the decision variables and problems concerning the local extrema. Problems connected with finding the global minimum of objective function occur very often in the case of design optimisation of many mechatronic systems [11]. Moreover, the choice of reliable and efficient optimisation criteria causes many difficulties for the designers. Some problems concerning the optimisation of pneumatic vibration isolators for various optimisation criteria are shown in the paper [12]. There is a lack of effective procedures of formulating and solving the optimisation problems for vibration reduction systems that are described by non-linear equations of motion. In addition, it is not known how to formulate a global objective function and which of the optimisation methods is effective for a given problem [13].

The second group of vibration reduction methods comprises the active techniques where the structural and parametric modifications are applied with the use of external power supply [14]. The actuator used in the system may generate a force that compensates the vibration coming from the exciting forces (active systems) or may change the system parameters (semi-active systems). The models of such vibration reduction systems are shown in Fig. 1.5.

The active vibration reduction systems generate the additional forces or controlled displacements which are operating directly on the isolated body so that to compensate the exciting forces or kinematic excitations (Fig. 1.5c). Unfortunately, the main drawback of such systems is the fact that they require a high amount of the power supply, and therefore, the active vibration isolators are hardly ever used in practice. Semi-active systems consist of passive elements (springs and dampers), although their force characteristics can be changed very fast (Fig. 1.5a–b). Such systems do not require strong power supply. Thus, semi-active vibration isolators are very often used in practice [15, 16].

The main problem of the semi-active or active vibration isolators concerns the selection of control strategy. Classical methods of vibration control significantly improve the vibro-isolation properties of the linear systems while the system parameters are exactly known. Unfortunately, the systems discussed in the literature are based on the simplified linear models that result in significant discrepancies compared with the real vibration isolators. In reality, the system parameters vary in a wide range and this leads to the lower effectiveness of active vibration isolator. Therefore, the parameter variations have to be taken into account especially using simplified models describing the vibration reduction systems. In the paper [17], the robust control system of active car suspension is presented that takes into account the uncertainties of visco-elastic elements used in the system. Successively in the paper [18], the robust control strategy is applied to vibration control of a rotor that

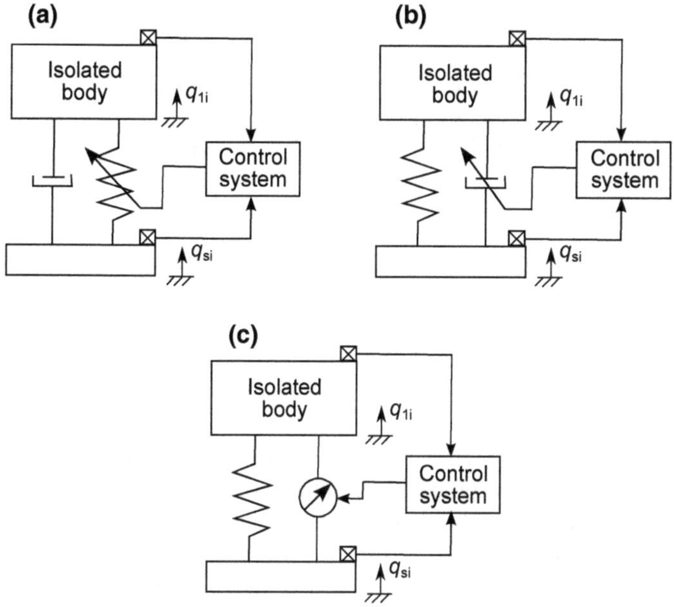

Fig. 1.5 Models of the vibration reduction systems: semi-active system with spring control (**a**), semi-active system with damper control (**b**), active system (**c**)

is suspended using the magnetic bearings with non-linear characteristics. In many papers [19–21], the modern vibration isolators are proposed and these systems are still being developed in terms of the effective control strategy.

In the recent literature, there is no systematic knowledge concerning the procedures to be followed in the design process of semi-active or active vibration reduction systems. The control system synthesis, especially if this synthesis is based on non-linear models, is difficult, when the semi-active or active system has to restrict conflicted vibro-isolation criteria. It is also not clearly defined, which of the controllers [22–25] can be applied in modern vibration isolator and how their controller settings should be calculated. Summarising, there is no effective method to define the control system structure and optimise the controller settings that can be utilised for the designed semi-active or active vibration reduction system.

1.2 Research Methodology

The fundamental method of evaluating the effectiveness of vibration reduction system is to perform an experiment in the laboratory. The tested system should be excited by the signals that are representative of the different types of working machines [26–28]. The vibro-isolation criteria of the tested system can be calculated on the

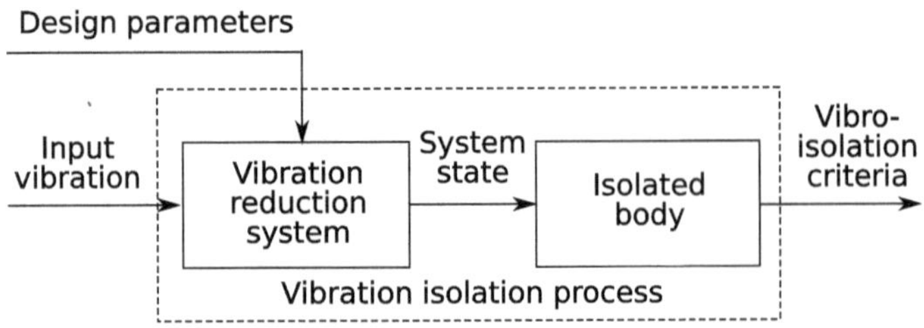

Fig. 1.6 Concept model of the vibro-isolation process

basis of measured signals. However, in many cases the technological restrictions and the high cost of the experiment make the test difficult to perform. The duration of the test is also an important aspect, especially when the test must be performed repeatedly for different design parameters of the system. Taking into account the complexity of the research process, the authors of this book recommend to carry out a simulation experiment based on a mathematical model of vibration isolation system that is shown in Fig. 1.6.

In order to analyse the dynamics of such system (Fig. 1.6), at first a theoretical model of the excitation signal with specific spectral characteristics should be elaborated. Subsequently, in order to determine the values of vibro-isolation criteria, which are very often defined as the integral values [14, 15], models of the vibration reduction system and the isolated body have to be created. For the purpose of this book, the dynamic behaviour of the vibration reduction systems is modelled in the MATLAB-Simulink® software package. The equations of motion are programmed using the interactive graphical environment, which allows to simulate and test a variety of the time-varying system.

The research presented in the papers [3, 6, 7] clearly shows that the passive vibration isolation methods in many cases prove to be ineffective. Although they provide energy dissipation at the sufficiently high frequencies, the low-frequency vibrations are amplified due to the resonance effect. In the consequence of this undesirable effect, it is difficult to achieve the desired system properties in order to meet the conflicted requirements for modern vibration reduction systems. The semi-active or active systems should improve the effectiveness of vibration isolators, and their concept model is shown in Fig. 1.7.

The feedback control of the vibration reduction system is intended to improve the dynamics of isolated body and shape the system characteristics especially for a specific vibro-isolation process realised by the system [2]. The control element in semi-active and active vibration isolation systems is adjusted by means of the controller which uses information concerning the system state and coming from a measurement system [1, 6]. However, the effective control system should allow to shape vibro-isolation properties of the system for the specified excitation signals and different working conditions. Using a unique vibration control system, whose structure and individual components will be proposed in this book, it will be possible

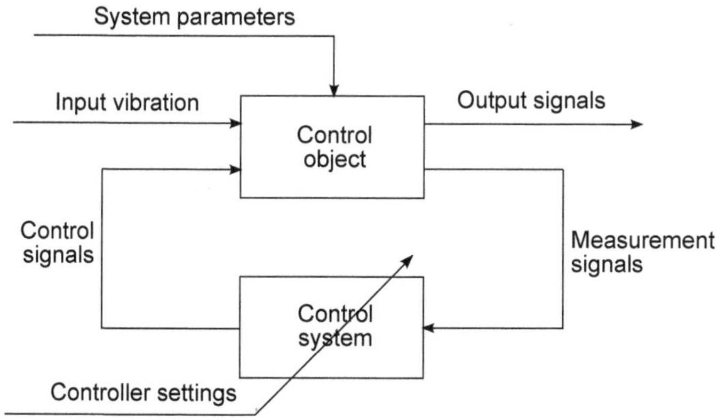

Fig. 1.7 Concept model of the semi-active or active vibration isolation system

to achieve the desired system properties in view of the conflicted requirements for modern vibration reduction systems.

The optimisation procedure proposed in this book ensures finding a set of the compromise solutions (Pareto-optimal solutions). This procedure allows to adjust the vibro-isolation properties of vibration isolator for the individual requirements defined by the machine operator. The required configurability of passive systems will be obtained by properly selecting their non-linear visco-elastic characteristics or by suitable changing of the controller settings (decision variables) in semi-active and active systems.

A selection of the vibro-isolation properties for the exemplary systems has to be performed within the framework of this work. Such a selection of the dynamic characteristics is conducted for the vibration isolation systems of different designs that are generated using the signals representing the work of different machinery. The optimisation process is led using the randomly starting points, because such procedure ensures a high probability that the optimum found is a global one. Initially, the separate minimising of the conflicted vibro-isolation criteria is carried out, and then, a minimisation of the primary criterion is recommended taking into account the other optimisation criteria.

References

1. Engel, Z., & Kowal, J. (1995). *Control of vibroacoustic processes (in Polish)*. Cracow: AGH University of Science and Technology Press.
2. Preumont, A. (2002). *Vibration Control of Active Structures An Introduction*. London: Kluwer Academic Publishers.
3. Engel, Z. (1993). *Environmental protection against vibration and sound*. Warsaw: PWN.
4. Griffin, M. J., Howarth, H. V. C., Pitts, P. M., Fischer, S., Kaulbars, U., & Donati, P. M., et al. (2006). *Guide to good practice on whole-body vibration. Non-binding guide to good*

practice with a view to implementation of Directive 2002/44/EC on the minimum health and safety requirements regarding the exposure of workers to the risks arising from physical agents (vibrations). Luxembourg: European Commission.
5. International Organization for Standardization. (1997). *Mechanical vibration and shock - Evolution of human exposure to whole body vibration*, ISO 2631, Genewa.
6. Kowal, J. (1996). *Vibration control (in Polish)*. Cracow: Gutenberg.
7. Nader, M. (2001). *Modelling and simulation of vehicle vibration influence on human body (in Polish)*. Warsaw: Publishing House of Warsaw University of Technology.
8. Cempel, C., Barczewski, R., & Dobry, M. W. (1984). Dynamics and vibroisolation efficiency of constant interaction force isolator model study. *Machine Dynamics Problems*, *2*(3), 135–150.
9. Maciejewski, I. (2005). *Research into the effectiveness of pneumatic vibration isolation systems for health protection of working machines operators against vibrations*, PhD thesis, Mechanical Department of Koszalin University of Technology, Koszalin.
10. Alkhatiba, R., Nakhaie, J. G., & Golnaraghi, M. F. (2004). Optimal design of passive linear suspension using genetic algorithm. *Journal of Sound and Vibration*, *275*, 665–691.
11. Tarnowski, W. (2009). *Optimisation and polyoptimisation in mechatronics (in Polish)*. Koszalin: Academic Publishing of the Koszalin University of Technology.
12. Kiczkowiak, T., & Maciejewski, I. (2010). Pneumatic vibro-isolation system - problems of quality assessment (in Polish). *Pneumatics*, *1*, 23–26.
13. Marler, R. T., & Arora, J. S. (2004). Survey of multi-objective optimization methods for engineering. *Structural and Multidisciplinary Optimization*, *26*, 369–395.
14. Coodall, R. M., & Kortm, W. (1983). Active control in ground transportation - review of the state-of-the-art and future potential. *Vehicle System Dynamics*, *12*, 225–257.
15. Choi, S. B., Lee, H. K., & Hang, E. G. (2001). Field test results of a semi-active ER suspension system associated with skyhook controller. *Mechatronics*, *11*, 345–353.
16. Khajavi, M. N., & Abdollahi, M. N. (2007) Comparison between optimized passive vehicle suspension system and semi active fuzzy logic controlled suspension system regarding ride and handling. In *Proceedings of World Academy of Science, Engineering and Technology* (Vol. 21).
17. Gu, D., Petkov, P., & Konstantinov, M. (2005). *Robust control design with MATLAB*. Berlin: Springer-Verlag.
18. Gosiewski, Z., & Mystkowski, A. (2006). The robust control of magnetic bearings for rotating machinery. *Solid State Phenomena*, *113*, 125–130.
19. Neubauer, M., Olekiewicz, R., Popp, K., & Krzyynski, T. (2006). Optimization of damping and absorbing performance of shunted piezoelements utilizing negative capacitance. *Journal of Sound and Vibration*, *298*, 84–107.
20. Maslanka, M., & Sapiski, B. (2007). Experimental study of vibration control of a cable with an attached MR damper. *Journal of Theoritical and Applied Mechanics*, *45*(4), 893–917.
21. Makowski, M., & Grzesikiewicz, W. (2008). Modeling of vibration of mechanical system with control dampers (in Polish). *Engineering modelling*, *36*, 221–228.
22. Ying, Z. G., Ni, Y. Q., & Ko, J. M. (2007). A bounded stochastic optimal semi-active control. *Journal of Sound and Vibration*, *304*, 948–956.
23. Poussot-Vassal, C., Sename, O., Dugard, L., Gaspar, P., Szabo, Z., & Bokor, J. (2008). A new semi-active suspension control strategy through LPV technique. *Control Engineering Practice*, *16*, 1519–1534.
24. Lauwerys, C., Swevers, J., & Sas, P. (2005). Robust linear control of an active suspension on a quarter car test-rig. *Control Engineering Practice*, *13*, 577–586.
25. Gaoa, H., Lam, J., & Wang, C. (2006). Multi-objective control of vehicle active suspension systems via load-dependent controllers. *Journal of Sound and Vibration*, *290*, 654–675.
26. British Standards Institution BS EN 13490. (2002). *Mechanical vibration - Industrial trucks - Laboratory evaluation and specification of operator seat vibration*, London.
27. International Organization for Standardization. (2003). *Agricultural wheeled tractors - Operator's seat - Laboratory measurement of transmitted vibration*, ISO 5007, Genewa.
28. International Organization for Standardization. (2000). *Earth-moving machinery - Laboratory evaluation of operator seat vibration*, ISO 7096, Genewa.

Chapter 2
Simulated Input Vibration

2.1 Nomenclature

$D_{si}(s)$ denominator polynomials of a total filter
$G_{si}(s)$ transfer function of a total filter
$G_{HPij}(s)$ transfer function of the Butterworth high-pass filter
$G_{LPij}(s)$ transfer function of the Butterworth low-pass filter
LSE_{si} least square error of a specified frequency response
$N_{si}(s)$ numerator polynomials of a total filter
PSD_{si} desired power spectral density, $(m/s^2)^2/Hz$
$P\hat{S}D_{si}$ estimated power spectral density, $(m/s^2)^2/Hz$
$P\bar{S}D_{si}$ mean value of the desired power spectral density, $(m/s^2)^2/Hz$
R_{si} multiple correlation coefficient
$a_1, ..., a_n$ coefficients of the Butterworth filter
f frequency, Hz
f_{HPij} cut-off frequency of the high-pass filter, Hz
f_{LPij} cut-off frequency of the low-pass filter, Hz
i directions of the vibration exposure (x, y, z)
j number of the signal generator (1, ...,l)
k_{ij} gain factor of the required magnitude of the power spectral density function in a specified frequency bandwidth, $(m/s^2)^2/Hz$
n filter order
$p(\ddot{q}_{ij}(t))$ probability density function of the acceleration signal
$\ddot{q}_{ij}(t)$ acceleration of the input vibration, m/s^2
$(\ddot{q}_{si})_{RMS}$ root mean square value of the acceleration signal, m/s^2
s Laplace variable
t_k computation time, s
t_s time interval between samples, s
σ_{ij}^2 variance of the random numbers.

T. Krzyzynski et al., *Modelling and Control Design of Vibration Reduction Systems*, https://doi.org/10.1007/978-3-030-03047-6_2

2.2 Introduction

Machine operators during their work are exposed to vibrations that are very often caused by the movement of machinery over uneven ground. In the case of whole-body vibration, the seat of a vehicle or the platform of a worker vibrates and this vibration is transmitted to the human body. Under normal operating conditions, the humans occupy the following body positions [1, 2]:

- sitting position (mechanical vibrations transmitted through the pelvis) that precludes active damping of vibration using the lower limbs,
- standing position (mechanical vibrations transmitted through the feet) that allows active damping of vibration in the low-frequency range.

The basic positions of the human body at work are presented in Fig. 2.1. Harmful vibrations can be transmitted to the human body in the three orthogonal directions: longitudinal x, lateral y and vertical z.

Vibrations are defined by their magnitudes and frequencies. The magnitudes of vibration are traditionally expressed as the vibration acceleration (in m/s^2), because most vibration transducers produce an output that is related to acceleration signal. The frequencies appearing in random whole-body vibration environment usually occur between 1 and 20 Hz [3]. In the following chapter, the theoretical model of input vibration is developed for different types of working machines. At first, a signal generator is utilised to generate random vibration. Then, a specific signal processing technique is used in order to obtain the spectral characteristics that are representative for the different types of working machine.

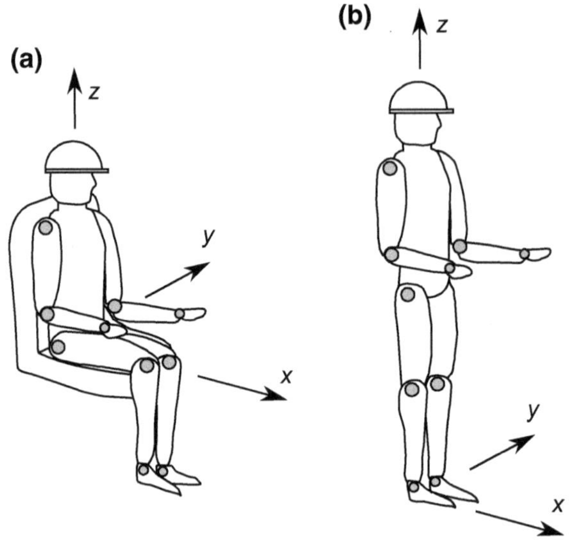

Fig. 2.1 Positions of the human body at work: sitting position (**a**) and standing position (**b**)

2.3 Random Signal Generator

The simulated input vibration is generated using a normally (Gaussian) distributed random numbers (rand) that produces such a waveform:

$$\ddot{q}_{ij}(t) = \sqrt{\sigma_{ij}^2}\text{rand}\left(\frac{t_k}{t_s}\right), \quad i = x, y, z, \quad j = 1, ..., l \tag{2.1}$$

where $\ddot{q}_{ij}(t)$ is the vibration acceleration generated for the direction x, y, z, $j = 1, ..., l$ is the number of signal generator, σ_{ij}^2 is the variance of the random numbers, t is the present moment of time, t_k is the computation time, and t_s is the time interval between samples. For accurate simulations, a sampling time should be much smaller than the fastest dynamics of the system. Acceptable results can be achieved by specifying $t_s = 0, 01/f_{max}$, where f_{max} is the bandwidth of the system in Hz.

The probability density function of normally distributed random signal can be described using the following relation:

$$p(\ddot{q}_{ij}(t)) = \frac{1}{\sqrt{2\pi\sigma_{ij}^2}}\exp\left(-\frac{1}{2\sigma_{ij}^2}\ddot{q}_{ij}^2(t)\right), \quad i = x, y, z, \quad j = 1, ..., l \tag{2.2}$$

The time plot of such defined stochastic signal $\ddot{q}_{ij}(t)$ is shown in Fig. 2.2a, and its histogram on the background of the probability density function is illustrated in Fig. 2.2b.

The spectral properties of the generated random signal have to be close to the white noise, so the power spectral density (PSD) should be flat in the considered frequency range and the correlation time of a time series shall be close to zero [4]. The power spectral density and the normalised autocorrelation of generated signal are presented in Fig. 2.3.

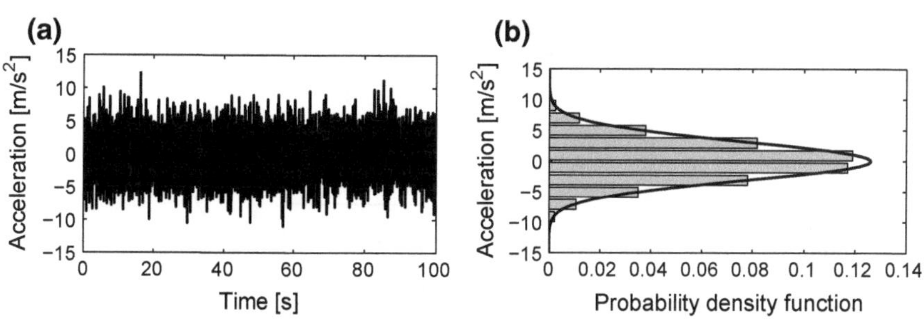

Fig. 2.2 Input vibration generated as normally distributed random process (**a**) and its histogram (■) on the background of the probability density function (—) (**b**)

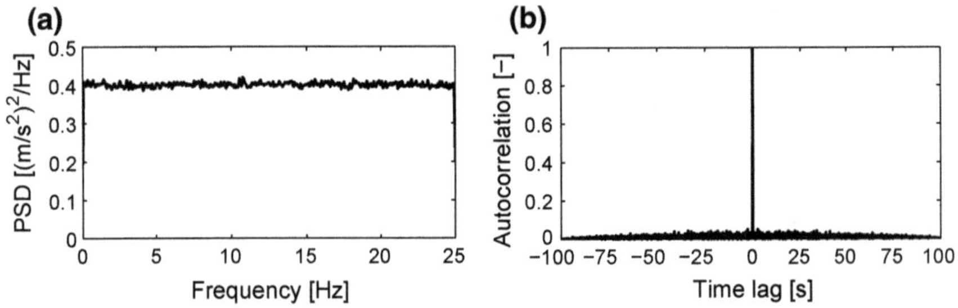

Fig. 2.3 Power spectral density (**a**) and normalised autocorrelation (**b**) of the generated random signal

2.4 Signal Processing Technique

An original signal processing technique is proposed in order to obtain specific spectral properties of the generated random signal. This technique involves the use of a set of the Butterworth filters (high pass and low pass). The block diagram of the proposed signal processing technique is shown in Fig. 2.4.

Transfer functions for the linear Butterworth filters are defined in the following way [5]:

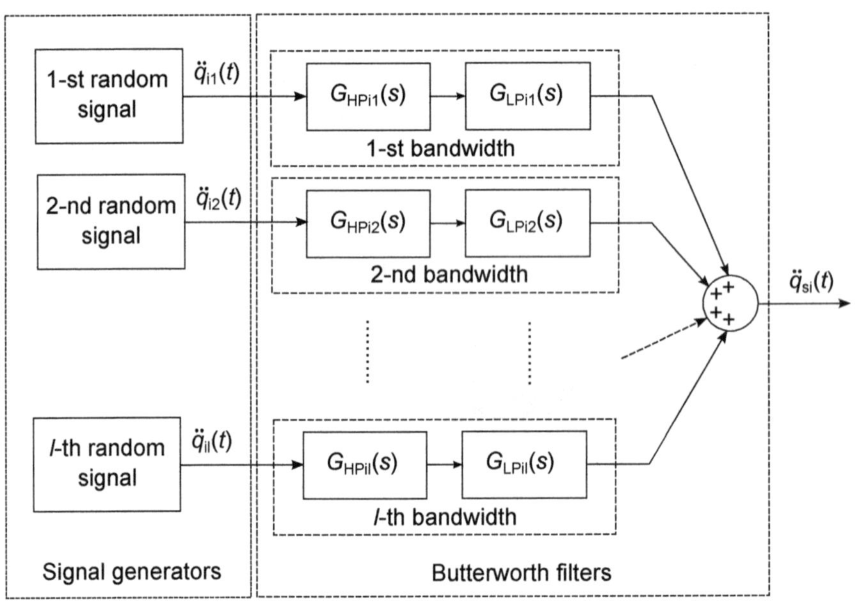

Fig. 2.4 Block diagram of the proposed signal processing technique

- high-pass filter

$$G_{\mathrm{HPij}}(s) = \frac{s^{\mathrm{n}}}{s^{\mathrm{n}} + a_{\mathrm{n}-1}s^{\mathrm{n}-1} + \cdots + a_1 s + 1}, \quad \mathrm{i} = x, y, z, \quad \mathrm{j} = 1, \ldots, l \tag{2.3}$$

- low-pass filter

$$G_{\mathrm{LPij}}(s) = \frac{1}{s^{\mathrm{n}} + a_{\mathrm{n}-1}s^{\mathrm{n}-1} + \cdots + a_1 s + 1}, \quad \mathrm{i} = x, y, z \quad \mathrm{j} = 1, \ldots, l \tag{2.4}$$

where a_1 to a_{n} are the coefficients of the Butterworth filter specified in Table 2.1, n is the order of the filter, s is the Laplace variable, $\mathrm{j} = 1, \ldots, l$ is the number of filters used.

The high-pass and low-pass filters (Eqs. 2.3 and 2.4) allow to create required band-pass filter for a specified frequency bandwidth. A combination of the high-pass and low-pass filters at the defined cut-off frequency and filter order provides generation of the input vibration with various spectral characteristics. The transfer function of a total filter is defined as follows:

$$G_{\mathrm{si}}(s) = \frac{N_{\mathrm{si}}(s)}{D_{\mathrm{si}}(s)} = \sum_{j=1}^{l} G_{\mathrm{HPij}}(s) G_{\mathrm{LPij}}(s), \quad \mathrm{i} = x, y, z, \quad \mathrm{j} = 1, \ldots, l \tag{2.5}$$

where $N_{\mathrm{si}}(s)$ and $D_{\mathrm{si}}(s)$ are the numerator and denominator polynomials that represent a combination of the Butterworth filters shown in Fig. 2.4.

The transfer function (Eq. 2.5) is especially useful when analysing the proposed filter stability. If all poles of this transfer function have negative real parts, then the filter is considered as asymptotically stable. However, the filter is unstable when any pole has a positive real part. Therefore, all poles on the complex s-plane should be located in the left half plane to ensure filter stability, which can be described by the following relation:

$$\mathrm{Re}\left(D_{\mathrm{si}}(s)\right) < 0, \qquad \mathrm{i} = x, y, z \tag{2.6}$$

Table 2.1 Coefficients of the Butterworth filter $a_1, \ldots, a_{\mathrm{n}}$ for different filter orders n

n	a_1	a_2	a_3	a_4	a_5	a_6	a_7
2	1,414	-	-	-	-	-	-
3	2,000	2,000	-	-	-	-	-
4	2,613	3,414	2,613	-	-	-	-
5	3,236	5,236	5,236	3,236	-	-	-
6	3,863	7,464	9,141	7,464	3,863	-	-
7	4,493	10,097	14,591	14,591	10,097	4,493	-
8	5,125	13,137	21,846	25,688	21,846	13,137	5,125

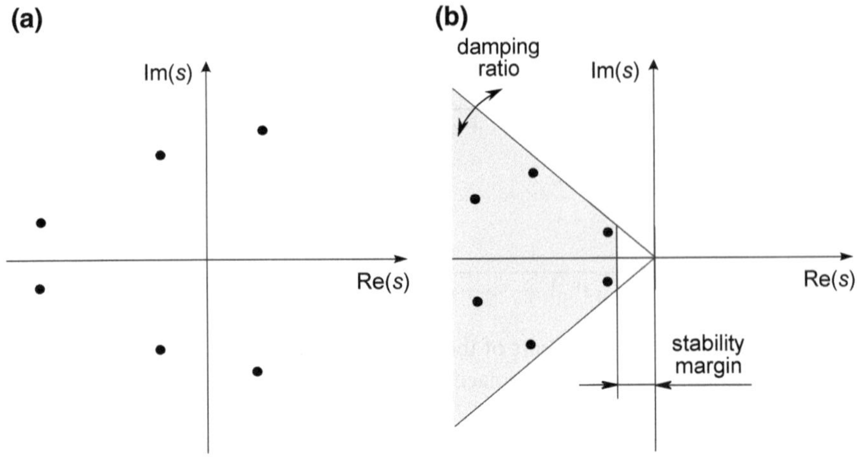

Fig. 2.5 Pole distribution on the complex s-plane: unstable filter (**a**) and stable filter with stability margin (**b**)

A pole distribution on the complex s-plane for unstable and stable filters is shown in Fig. 2.5.

The input vibration is defined by a power spectral density of the vertical acceleration and by root mean square value of the acceleration signal. The target magnitude $\mathrm{PSD_{si}}(2\pi f)$ of the power spectral density function can be calculated with the assistance of the following equation:

$$\mathrm{PSD_{si}}(2\pi f) = \sum_{j=1}^{l} \frac{2\sigma_{\mathrm{ij}}^2}{f_s} \left| \left(G_{\mathrm{HPij}}(2\pi f)\right)^2 \left(G_{\mathrm{LPij}}(2\pi f)\right)^2 \right|, \quad \begin{array}{l} \mathrm{i} = x, y, z, \\ \mathrm{j} = 1, \ldots, l \end{array} \tag{2.7}$$

where f is the frequency in Hz, σ_{ij}^2 is the signal variance in a specified frequency bandwidth, and f_s is the sampling frequency. The curve defined by Eq. (2.7) consists of target power spectral density to be produced for the simulated input vibration (Fig. 2.6).

The power spectral density of simulated acceleration signal is considered to be representative for the different types of working machine if, and only if [6–8]:

- the magnitude of simulated input is within the tolerance of the target power spectral density function: $\mathrm{PSD_{si}}(2\pi f) \pm 0.1 \max\limits_{f} (\mathrm{PSD_{si}}(2\pi f))$.
- the root mean square (RMS) value of simulated input acceleration is within the tolerance of the required value: $(\ddot{q}_{\mathrm{si}})_{\mathrm{RMS}} \pm 0.05\,(\ddot{q}_{\mathrm{si}})_{\mathrm{RMS}}$.

The root mean square value of the acceleration signal is defined using the following relation:

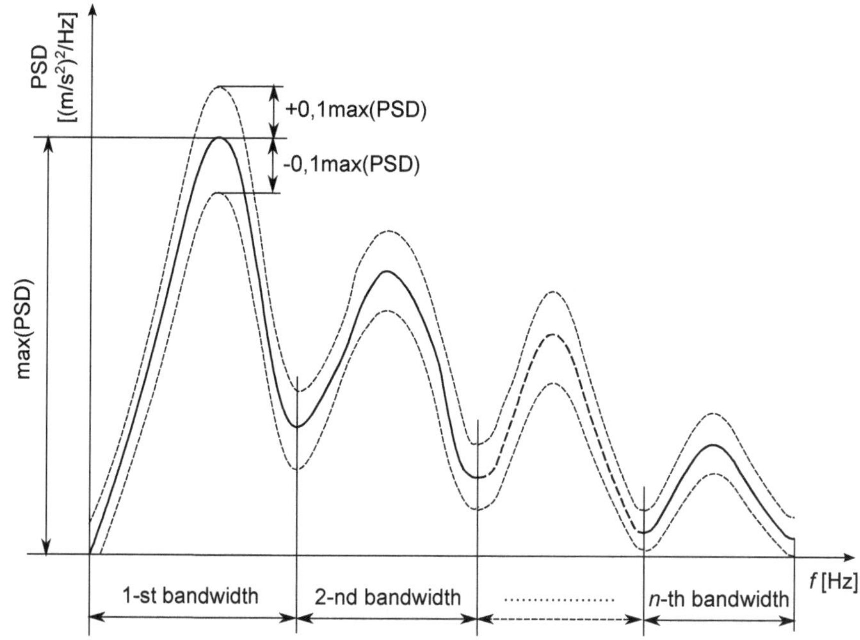

Fig. 2.6 Target power spectral density of the simulated input vibration

$$(\ddot{q}_{\mathrm{si}})_{\mathrm{RMS}} = \sqrt{\frac{1}{t_{\mathrm{k}}}\int_{0}^{t_{\mathrm{k}}}(\ddot{q}_{\mathrm{si}}(t))^2\,dt}, \qquad \mathrm{i} = x, y, z \tag{2.8}$$

where $\ddot{q}_{\mathrm{si}}(t)$ is the time history of input vibration and t_{k} is the time duration in which vibration data for the analysis are obtained.

2.5 Spectral Estimation Method

A novel method is proposed in which the PSD is estimated from a signal that is assumed to be the output of a linear system driven by a white noise. This method estimates the PSD by evaluating the parameters (coefficients) of the system shown in Fig. 2.4 that hypothetically generates the signal. Such a parametric method allows to find a multi-band filter with magnitude response approximating a desired function (Fig. 2.7).

The proposed method applies the least square error (LSE) minimisation technique over the frequency range of the filter's response. The least square error of a specified frequency response is described in the following form:

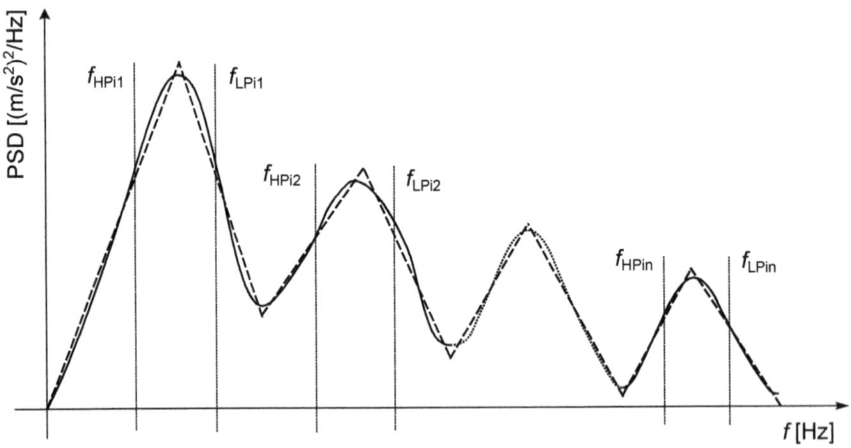

Fig. 2.7 Magnitude response (—) approximating a desired function (····)

$$\mathrm{LSE_{si}} = \sqrt{\sum_{k=1}^{m}\left(\mathrm{PSD_{si}}(2\pi f_k) - \hat{\mathrm{PSD}}_{\mathrm{si}}(2\pi f_k)\right)^2}, \quad \mathrm{i} = x, y, z \tag{2.9}$$

where $\mathrm{PSD_{si}}$ and $\hat{\mathrm{PSD}}$ are the desired and estimated power spectral densities that are obtained for the same frequency value f_k.

The error minimisation includes the design parameters having an influence on the spectral characteristics of the generated signal as follows:

$$\min_{\sigma_{\mathrm{ij}}^2,\, f_{\mathrm{HPij}},\, f_{\mathrm{LPij}}} \mathrm{LSE_{si}}\left(\sigma_{\mathrm{ij}}^2,\, f_{\mathrm{HPij}},\, f_{\mathrm{LPij}}\right), \quad \mathrm{i} = x, y, z, \quad \mathrm{j} = 1, \ldots, l \tag{2.10}$$

where σ_{ij}^2 is the variance of the random signal, f_{HPij} and f_{LPij} are the cut-off frequencies of the high-pass and low-pass Butterworth filters, respectively.

The problem of minimising the objective function (Eq. 2.10) of several parameters (decision variables) is subjected to linear inequality constraints on these variables:

$$\begin{array}{l} (\sigma_{\mathrm{ij}}^2)_{\min} \le \sigma_{\mathrm{ij}}^2 \le (\sigma_{\mathrm{ij}}^2)_{\max} \\ \left(f_{\mathrm{HPij}}\right)_{\min} \le f_{\mathrm{HPij}} \le \left(f_{\mathrm{HPij}}\right)_{\max} \\ \left(f_{\mathrm{LPij}}\right)_{\min} \le f_{\mathrm{LPij}} \le \left(f_{\mathrm{LPij}}\right)_{\max} \end{array} \quad \mathrm{i} = x, y, z, \quad \mathrm{j} = 1, \ldots, l \tag{2.11}$$

where $(\sigma_{\mathrm{ij}}^2)_{\min}$ and $(\sigma_{\mathrm{ij}}^2)_{\max}$ are the lowest and highest values of signal variance, $\left(f_{\mathrm{HPij}}\right)_{\min}$ and $\left(f_{\mathrm{HPij}}\right)_{\max}$ are the lowest and highest values of the cut-off frequencies of high-pass filter, and $\left(f_{\mathrm{LPij}}\right)_{\min}$ and $\left(f_{\mathrm{LPij}}\right)_{\max}$ are the lowest and highest values of the cut-off frequencies of low-pass filter.

Additionally, non-linear inequality constraints are imposed on the objective function (Eq. 2.10) to restrict the filter cut-off frequencies in the following order:

$$f_{\mathrm{HPi1}} \leq f_{\mathrm{LPi1}} \leq f_{\mathrm{HPi2}} \leq f_{\mathrm{LPi2}}, \ldots, \leq f_{\mathrm{HPij}} \leq f_{\mathrm{LPij}}, \quad \mathrm{i} = x, y, z, \quad \mathrm{j} = 1, \ldots, l \tag{2.12}$$

where $\mathrm{j} = 1, \ldots, l$ is the number of low-pass (LP) and high-pass (HP) filters used to estimate the input vibration in different directions ($\mathrm{i} = x, y, z$).

The multiple correlation coefficient (R) is employed in order to measure how successful the approximation is in explaining the variation of the optimisation results. This coefficient is defined as follows:

$$\mathrm{R}_{\mathrm{si}} = \sqrt{1 - \frac{\sum_{k=1}^{m}\left(\hat{\mathrm{PSD}}_{\mathrm{si}}(2\pi f_{\mathrm{k}}) - \mathrm{PSD}_{\mathrm{si}}(2\pi f_{\mathrm{k}})\right)^2}{\sum_{k=1}^{m}\left(\mathrm{PSD}_{\mathrm{si}}(2\pi f_{\mathrm{k}}) - \bar{\mathrm{PSD}}_{\mathrm{si}}(2\pi f_{\mathrm{k}})\right)^2}}, \quad \mathrm{i} = x, y, z \tag{2.13}$$

where $\hat{\mathrm{PSD}}_{\mathrm{si}}$ is the estimated power spectral density for the direction x, y, z, $\mathrm{PSD}_{\mathrm{si}}$ is the desired power spectral density, and $\bar{\mathrm{PSD}}_{\mathrm{si}}$ is the mean value of the desired power spectral density, f_{k} is the discrete value of frequency. The coefficient R_{si} can take on any value between 0 and 1, with a value closer to 1 indicating a better estimation of the power spectral density.

2.6 Vertical Vibration

The standards [6–8] specify the laboratory-simulated vertical vibration (z-axis) that is based on representative measured data from the different types of machines in typical working conditions. The input spectral classes are defined for the machines having similar mechanical characteristics. The test inputs indicated in these standards are based on a large number of measurements taken in situ of working machines while they were used under severe operating conditions.

The International Standard ISO 7096 [8] specifies the input vibrations for earth-moving machinery in nine spectral classes. The particular input vibration is defined as follows:

- EM1—articulated or rigid frame dumper >4500 kg,
- EM2—scraper without axle or frame suspension,
- EM3—wheel loader >4500 kg,
- EM4—grader,
- EM5—wheel dozer, sol compactor on wheels, backhoe loader,
- EM6—crawler loader, crawler dozer ≤50000 kg,
- EM7—compact dumper ≤4500 kg,
- EM8—compact loader ≤4500 kg,
- EM9—compact loader ≤4500 kg (skid steer type).

The British Standard BS EN 13490 [6] defines the input spectral classes required for the following industrial trucks (each class defines a group of machines having similar vibration characteristics):

- IT1—platform trucks, trucks rider-controlled with wheel mean diameter below 200 mm and high-load non-rubber solid tyres (category 1 according to [9]), reach trucks, articulated trucks, etc., with wheel mean diameter below 450 mm and high-load non-rubber solid tyres or cylindrical/conical base rubber solid tyres (category 2 according to [9]), straddle trucks, trucks with wheel mean diameter below 645 mm and rubber solid or pneumatic tyres (category 3 according to [9]),
- IT2—straddle trucks, trucks with wheel mean diameter between 645 mm and 900 mm and rubber solid or pneumatic tyres (category 4a according to [9]),
- IT3—straddle trucks, trucks with wheel mean diameter between 900 mm and 1200 mm and rubber solid or pneumatic tyres (category 4b according to [9]), trucks with wheel mean diameter between 1200 mm and 2000 mm and rubber solid or pneumatic tyres (category 5 according to [9]),
- IT4—all-terrain trucks (category 6 according to [9]).

The ISO 5007 standard [7] specifies the input vibration in three input spectral classes for agricultural tractors with rubber tyres, unsprung rear axles and no low-frequency cab isolation:

- AG1—agricultural tractors up to 3600 kg unballasted mass,
- AG2—agricultural tractors in the range of 3600 to 6500 kg unballasted mass,
- AG3—agricultural tractors over 6500 kg unballasted mass.

The standards [6–8] completely designate the high-pass and low-pass filters of the Butterworth type that are required for generating vibration along the vertical axis. Numerical values of the cut-off frequencies and filter orders are presented in Table 2.2. The vibration characteristics for each input spectral classes, i.e. EM1 through EM9, IT1 through IT4, AG1 through AG3 and their tolerances, are shown in Figs. 2.8 and 2.9. The desired and obtained root mean square values of the acceleration signal are defined in Table 2.2.

2.7 Horizontal Vibration

In contrary to vertical vibration, there are no standardised signals available for the horizontal directions (x-axis and y-axis). Therefore, this section specifies input vibrations for eight spectral classes and each class consisting of a longitudinal and lateral cabin floor vibration for particular working machines performing a specific operation. In the following paper [10], the test inputs were measured as a part of the work that has been done by Isringhausen GmbH & CO. KG and determined within the framework of the European research project VIBSEAT. They are not of sufficient various magnitudes to cover the majority of actual spectra observable on the field. The determination of representative spectra represents a large amount of work that is not the purpose of this book.

The test laboratory of Isringhausen GmbH & CO. KG presents the input vibrations in horizontal directions as follows:

Table 2.2 Parameters of the vertical input vibration generated for different types of working machines

			Signal generators			High-pass filters						Low-pass filters						Results	
			$\sigma_{\ddot{i}j}^{2}, \left(\frac{m}{s^2}\right)^2$			f_{HPij}, Hz			n_{HPij}			f_{LPij}, Hz			n_{LPij}			$(\ddot{q}_{si})_{RMS}, \frac{m}{s^2}$	
Type of machines	Input vibration	i =	j = 1	2	3	j = 1	2	3	j = 1	2	3	j = 1	2	3	j = 1	2	3	Desired	Obtained
Earth-moving machinery	EM1	z	282	–	–	1,5	–	–	4	–	–	2,5	–	–	4	–	–	1,17	1,67
	EM2	z	272	–	–	1,5	–	–	4	–	–	3	–	–	4	–	–	2,05	2,02
	EM3	z	193	–	–	1,5	–	–	4	–	–	3	–	–	4	–	–	1,75	1,70
	EM4	z	60	–	–	1,5	–	–	4	–	–	3	–	–	4	–	–	0,96	0,95
	EM5	z	111	–	–	1,5	–	–	4	–	–	3,5	–	–	1	–	–	1,94	2,10
	EM6	z	79	–	–	6,5	–	–	2	–	–	9	–	–	2	–	–	1,65	1,70
	EM7	z	925	–	–	3	–	–	8	–	–	3,5	–	–	8	–	–	2,25	2,27
	EM8	z	145	–	–	3	–	–	4	–	–	3	–	–	2	–	–	1,05	1,05
	EM9	z	210	–	–	3,5	–	–	4	–	–	4	–	–	2	–	–	1,63	1,64
Industrial trucks	IT1	z	166	–	–	4,5	–	–	4	–	–	5	–	–	2	–	–	1,58	1,57
	IT2	z	145	–	–	3	–	–	4	–	–	3	–	–	2	–	–	1,05	1,05
	IT3	z	60	–	–	1,5	–	–	4	–	–	3	–		4			0,96	0,94
	IT4	z	164	–	–	1,5	–	–	4	–	–	3	–	–	4	–	–	1,59	1,57
Agricultural wheeled tractors	AG1	z	925	–	–	3	–	–	8	–	–	3,5	–	–	8	–	–	2,26	2,27
	AG2	z	722	–	–	2,1	–	–	8	–	–	2,6	–	–	8	–	–	1,94	1,91
	AG3	z	585	–	–	1,95	–	–	8	–	–	2,45	–	–	8	–	–	1,74	1,71

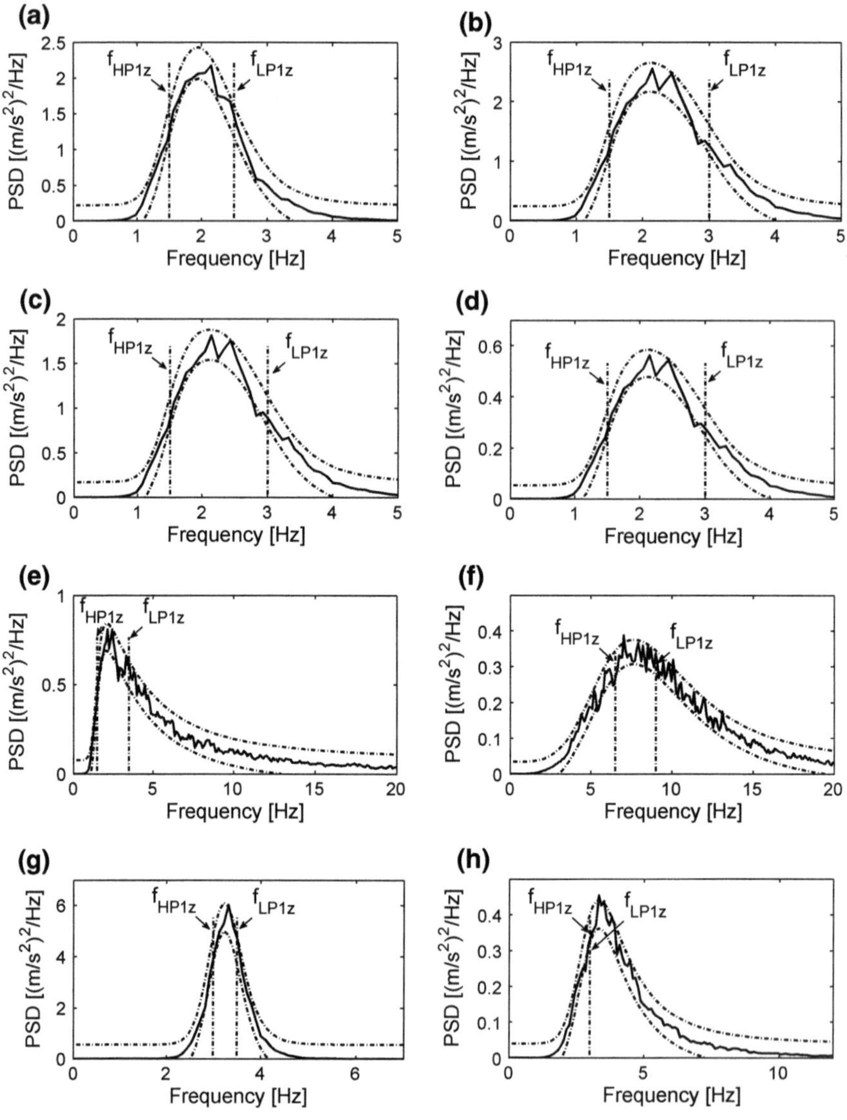

Fig. 2.8 Power spectral densities of simulated vertical input vibration (—) for the spectral classes: EM1 (**a**), EM2 (**b**), EM3 (**c**), EM4 (**d**), EM5 (**e**), EM6 (**f**), EM7 (**g**), EM8 (**h**) and their tolerances (-·-·-)

- AT1x—agricultural tractor on road at speed 30 kph (x-axis),
- AT1y—agricultural tractor on road at speed 30 kph (y-axis),
- AT2x—agricultural tractor on road at speed 40 kph (x-axis),
- AT2y—agricultural tractor on road at speed 30 kph (y-axis),
- WL1x—wheel loader with capacity of up to 15 tons (x-axis),
- WL2y—wheel loader with capacity of up to 15 tons (y-axis),

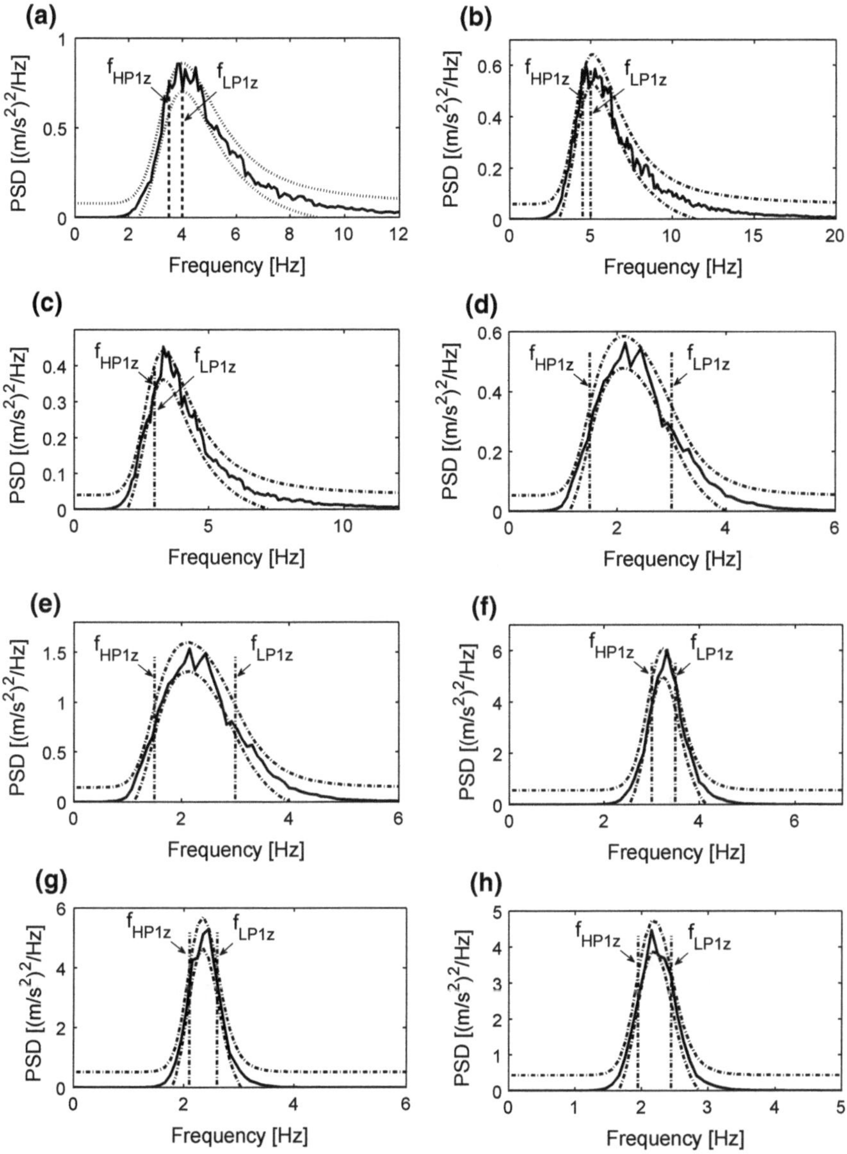

Fig. 2.9 Power spectral densities of simulated vertical input vibration (—) for the spectral classes: EM9 (**a**), IT1 (**b**), IT2 (**c**), IT3 (**d**), IT4 (**e**), AG1 (**f**), AG2 (**g**), AG3 (**h**) and their tolerances (-··-)

- WL1x—wheel loader with capacity of up to 30 tons (x-axis),
- WL2y—wheel loader with capacity of up to 30 tons (y-axis).

The test protocol within a work package of the VIBSEAT project recommends the following test inputs that are acquired under real field conditions:

Table 2.3 Parameters of the horizontal input vibration generated for different types of working machines

			Signal generators			High-pass filters						Low-pass filters						Results	
			$\sigma_{ij}^2, \left(\frac{m}{s^2}\right)^2$			f_{HPij}, Hz			n_{HPij}			f_{LPij}, Hz			n_{LPij}			$(\ddot{q}_{si})_{RMS}, \frac{m}{s^2}$	
Type of machines	Input vibration	i =	j = 1	2	3	j = 1	2	3	j = 1	2	3	j = 1	2	3	j = 1	2	3	Desired	Obtained
Agricultural wheeled tractors	AT1x	x	5,4	0,69	–	1,3	3,4	–	2	2	–	3	9,5	–	2	2	–	0,35	0,35
	AT1y	y	10,1	1,4	–	0,5	4,9	–	2	2	–	1,8	8,5	–	2	2	–	0,47	0,45
	AT2x	x	25	3,17	–	2	5,9	–	4	2	–	2,5	6,7	–	4	2	–	0,48	0,47
	AT2y	y	12,6	1,29	–	0,5	2,2	–	4	2	–	1,6	11	–	2	2	–	0,53	0,52
	AT3x	x	49	5,1	12	0,7	1,6	5,1	2	4	4	1,5	3,2	5,3	4	4	4	0,72	0,74
	AT3y	y	134	3,8	3,3	0,8	1,4	9,2	4	4	4	1,2	4,4	10	4	4	4	0,86	0,89
	AT4x	x	335	51	27,1	2	5,2	9,7	8	4	4	2,4	5,7	12	4	4	4	1,62	1,69
	AT4y	y	728	165	27,3	1,1	3	7,7	8	4	4	1,3	3,4	10	8	4	4	1,86	1,84
Wheel loaders	WL1x	x	996	91	–	1,2	2	–	8	4	–	1,4	3	–	8	4	–	2,06	1,97
	WL1y	y	231	200	–	1,2	1,8	–	8	4	–	1,4	2,2	–	8	4	–	1,36	1,29
	WL2x	x	438	350	–	0,7	4,2	–	8	4	–	1,8	4,4	–	4	8	–	2,81	2,61
	WL2y	y	178	105	–	0,8	4,3	–	8	4	–	2	5	–	4	8	–	2,04	1,92
Articulated truck	AL1x	x	7,9	6,8	3,7	0,2	1,3	5,7	8	4	4	1,2	2,3	7	8	4	2	0,50	0,49
	AL1y	y	15,7	21,9	1,8	0,4	2	6	4	8	4	1,5	2,4	14	4	8	2	0,62	0,63
Rail vehicle	RV1x	x	3,1	1,4	0,6	2,8	4,8	9	4	4	4	3	4,8	9,3	8	4	4	0,17	0,17
	RV1y	y	10,3	2,4	–	0,6	1,3	–	4	4	–	0,9	4,3	–	4	4	–	0,32	0,32

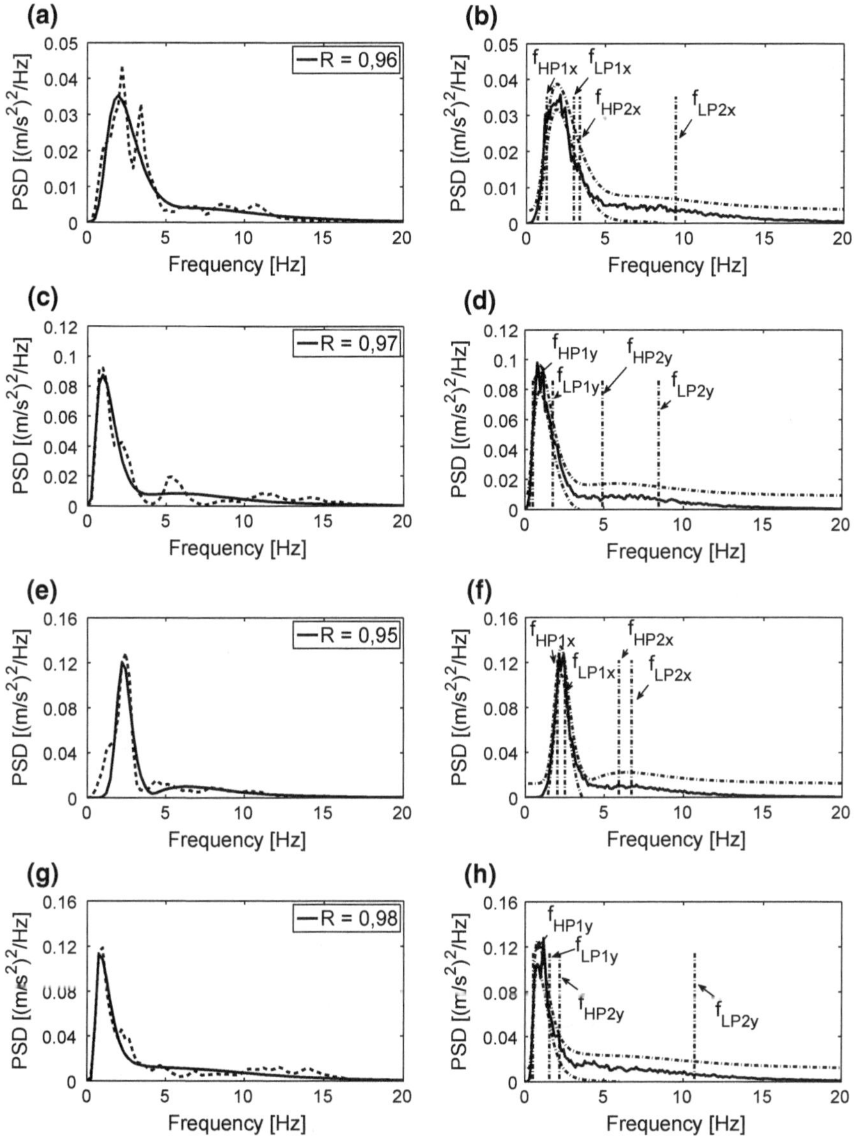

Fig. 2.10 Power spectral densities of measured (----), estimated (——) (left-hand side) and simulated (——) (right-hand side) horizontal input vibrations for the spectral classes: AT1x (**a**–**b**), AT1y (**c**–**d**), AT2x (**e**–**f**), AT2y (**g**–**h**) and their tolerances (-·-·-)

- AL1x—articulated truck over rough ground (x-axis),
- AL2y—articulated truck over rough ground (y-axis),
- RV1x—rail vehicle (x-axis),
- RV2y—rail vehicle (y-axis),
- AT3x—agricultural tractor at speed 3 kph (x-axis),

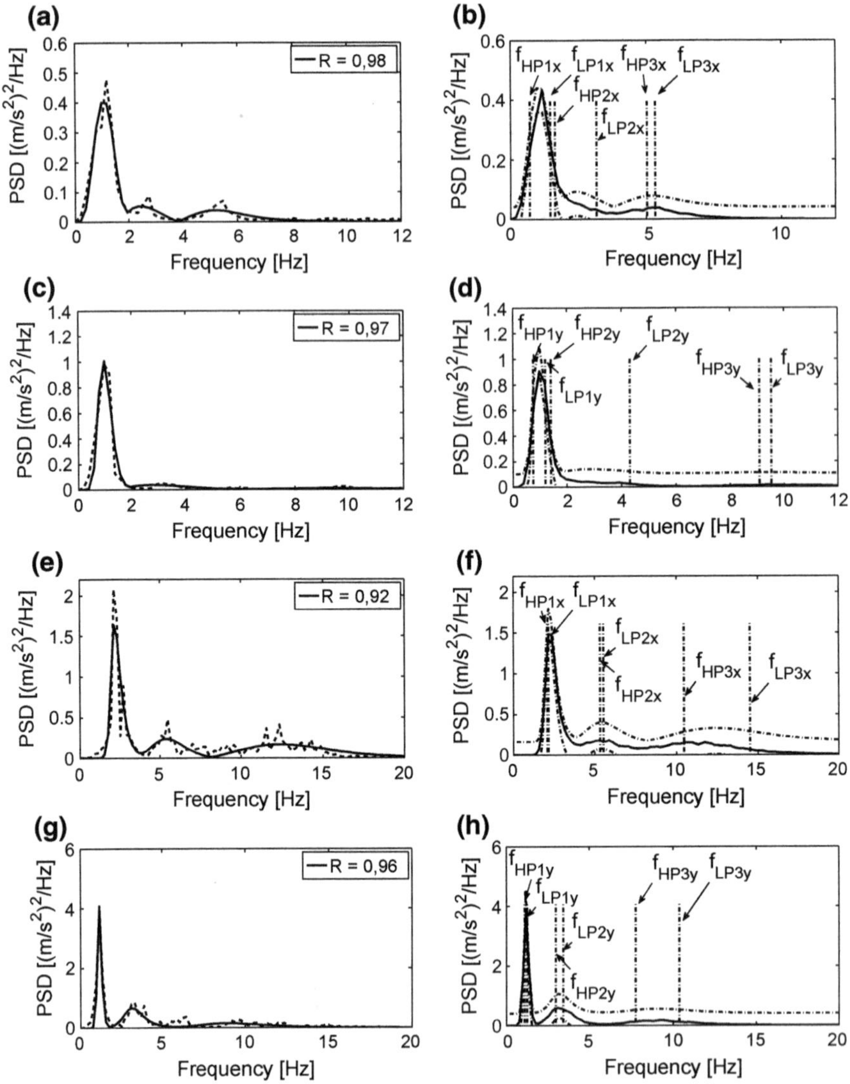

Fig. 2.11 Power spectral densities of measured (----), estimated (——) (left-hand side) and simulated (——) (right-hand side) horizontal input vibrations for the spectral classes: AT3x (**a**–**b**), AT3y (**c**–**d**), AT4x (**e**–**f**), AT4y (**g**–**h**) and their tolerances (-·-·-)

- AT3y—agricultural tractor at speed 3 kph (y-axis),
- AT4x—agricultural tractor at speed 20 kph (x-axis),
- AT4y—agricultural tractor at speed 20 kph (y-axis).

There are no designated Butterworth filters that are desired for generating vibration along the horizontal axes. Therefore, at first the PSD functions of these signals are determined using the spectral estimation method shown in Sect. 2.5. Then, the vari-

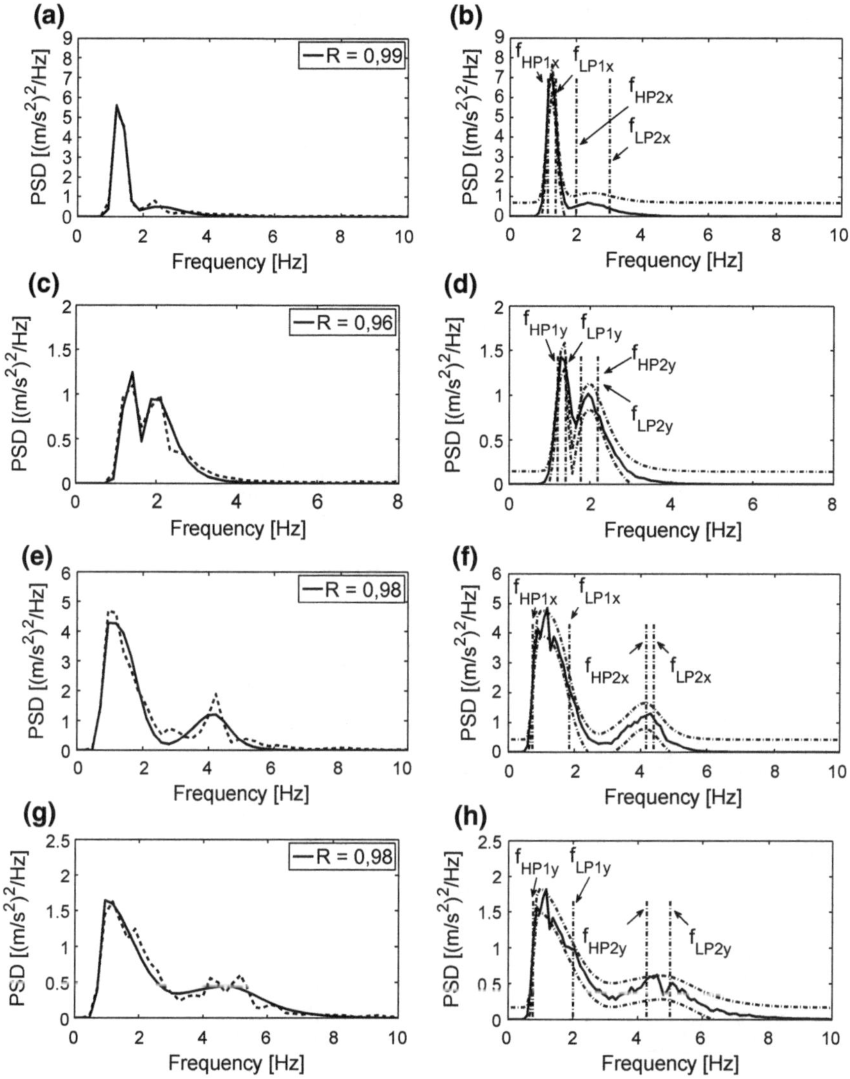

Fig. 2.12 Power spectral densities of measured (----), estimated (——) (left-hand side) and simulated (——) (right-hand side) horizontal input vibrations for the spectral classes: WL1x (**a**–**b**), WL1y (**c**–**d**), WL2x (**e**–**f**), WL2y (**g**–**h**) and their tolerances (-·-·-)

ance of random signals and the cut-off frequencies of filters are evaluated using the objective function described by Eq. (2.10). Numerical values of the design parameters are presented in Table 2.3. The spectral characteristics of measured, estimated and simulated input vibrations, i.e. AL1 through AL2, RV1 through RV2, AT3 through AT4 and their tolerances, are shown in Figs. 2.10, 2.11, 2.12 and 2.13. The desired and obtained root mean square values of the acceleration signal are defined in Table 2.3.

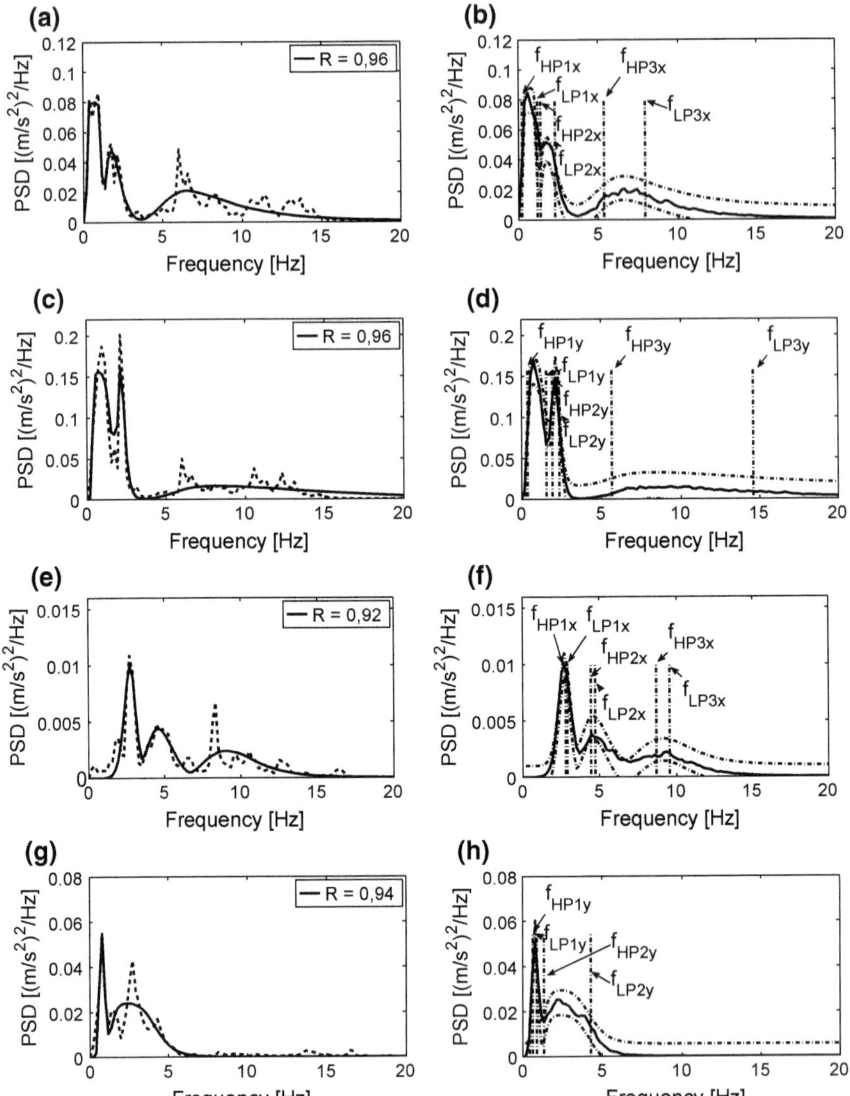

Fig. 2.13 Power spectral densities of measured (----), estimated (——) (left-hand side) and simulated (——) (right-hand side) horizontal input vibrations for the spectral classes: AL1x (**a**–**b**), AL1y (**c**–**d**), RV1x (**e**–**f**), RV1y (**g**–**h**) and their tolerances (-·-·-)

2.8 Conclusions

In this chapter, the theoretical models of simulated input vibrations are developed in such a way that the specific spectral properties of signals are obtained using the original filtration technique. This is the basis of the proposed procedure for generating random vibration that is representative for vibrations affecting the operators at work. Models developed in the framework of this chapter will be used to reproduce the real working conditions of vibration reduction systems in different types of machines. Investigations carried out in the next part of the following work assist a selection process of vibro-isolation properties of the systems for different spectral classes of the input vibration.

References

1. Griffin, M. J., Howarth, H. V. C., Pitts, P. M., Fischer, S., Kaulbars, U., & Donati, P. M., et al. (2006). *Guide to good practice on whole-body vibration. Non-binding guide to good practice with a view to implementation of Directive 2002/44/EC on the minimum health and safety requirements regarding the exposure of workers to the risks arising from physical agents (vibrations).* Luxembourg: European Commission.
2. Nader, M. (2001). *Modelling and simulation of vehicle vibration influence on human body (in Polish).* Warsaw: Publishing House of Warsaw University of Technology.
3. Maciejewski, I., Meyer, L., & Krzyzynski, T. (2009). Modelling and multi-criteria optimisation of passive seat suspension vibro-isolating properties. *Journal of Sound and Vibration, 324*, 520–538.
4. Bendat, J. S., & Piersol, A. G. (2004). *Methods for analysis and measurement of random signals (in Polish).* Warsaw: Polish Scientific Publishers PWN.
5. Parks, T. W., & Burrus, C. S. (1987). *Digital filter design.* New York: Wiley.
6. British Standards Institution BS EN 13490. (2002). *Mechanical vibration - Industrial trucks - Laboratory evaluation and specification of operator seat vibration*, London.
7. International Organization for Standardization. (2003). *Agricultural wheeled tractors - Operator's seat - Laboratory measurement of transmitted vibration*, ISO 5007, Genewa.
8. International Organization for Standardization. (2000). *Earth-moving machinery - Laboratory evaluation of operator seat vibration*, ISO 7096, Genewa.
9. International Organization for Standardization. (1987). *Powered industrial trucks - Terminology*, ISO 5053, Genewa.
10. Clement, P. (2005). *Laboratory test protocol for assessing the vibration performance of seats equipped with horizontal suspensions*, Technical Report for the Competitive and Sustainable Growth Programme. VIBSEAT G3RD-CT-2002-00827, No. 4 (pp. 1–26).

Chapter 3
Modelling of the Vibration Reduction System

3.1 Nomenclature

- $\mathbf{A}_{si}$ State (system) matrix (i = x, y, z)
- $\mathbf{B}_{si}$ Input vector (i = x, y, z)
- $\mathbf{C}_{si}$ Output matrix (i = x, y, z)
- $\mathbf{C}_{i}$ Stiffness matrix of the bio-mechanical model of human body (i = x, y, z)
- $\mathbf{D}_{si}$ Feedthrough (feedforward) vector (i = x, y, z)
- $\mathbf{D}_{i}$ Damping matrix of the bio-mechanical model of human body (i = x, y, z)
- F_{1i} Applied force of the human body model for selected direction of the vibration exposure (i = x, y, z), N
- $\mathbf{F}_{ai}$ Vector of the active forces (i = x, y, z)
- F_{aij} Applied force of the active element (i = x, y, z), N
- F_{cij} Applied force of the conservative element (i = x, y, z), N
- F_{dij} Applied force of the dissipative element (i = x, y, z), N
- $\mathbf{F}_{si}$ Vector of the exciting forces (i = x, y, z)
- $H_i(s)$ transfer function of the human body model for selected direction of the vibration exposure (i = x, y, z)
- $\mathbf{M}_{i}$ Inertia matrix of the bio-mechanical model of human body (i = x, y, z)
- M_i Apparent mass of the human body model for selected direction of the vibration exposure (i = x, y, z), kg
- f Frequency of vibration, Hz
- i Directions of the vibration exposure (x, y, z)
- q_{1i} Displacement of the isolated body for selected direction of the vibration exposure (i = x, y, z), m
- $\ddot{q}_{1i}$ Acceleration of the human body model for selected direction of the vibration exposure (i = x, y, z), $\mathrm{m/s^2}$
- $\mathbf{q}_{i}$ Displacement vector of the bio-mechanical model of human body (i = x, y, z)
- q_{si} Displacement of the input vibration (i = x, y, z), m
- $\mathbf{x}_{si}$ State vector (i = x, y, z).

T. Krzyzynski et al., *Modelling and Control Design of Vibration Reduction Systems*, https://doi.org/10.1007/978-3-030-03047-6_3

3.2 Introduction

The basic opportunity to minimise harmful vibrations consists in applying a vibration reduction system that prevents a propagation of the mechanical vibrations from their source to the isolated body. The dynamic characteristics of vibration isolator should be selected for a specific application in such a way that the vibrations transmitted from the source to the body are minimal.

In the following chapter, the generalised mathematical model of vibration isolation system is developed. The created model enables to shape non-linear characteristics of the visco-elastic elements included in vibration isolator. Such a mathematical description of the vibration reduction system determines the ability to use the developed model in a wide range of applications.

3.3 Generalised Model of the System

The generalised model of vibro-isolation system is presented in Fig. 3.1. The human body is protected against vibrations occurring along three mutually perpendicular axes, i.e. two horizontal (x-axis and y-axis) and one vertical (z-axis). In order to limit vibrations acting on the human body, passive visco-elastic elements are utilised in conjunction with active force actuators. Rotation vibrations are ignored due to simplifications that are formulated within the framework of this book. As follows from the European Directive [1], research on the harmful effects of vibrations should

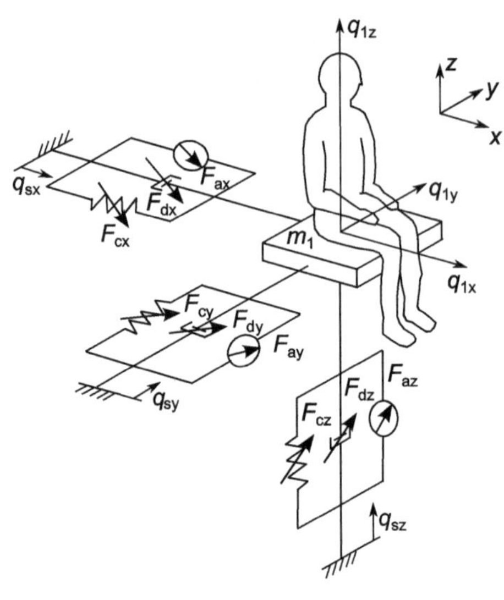

Fig. 3.1 Generalised model of the non-linear vibration reduction system

be performed only for the translational axes (x, y, z) neglecting rotations around each axis.

A set of three independent equations of motion is formulated in the matrix form:

$$\mathbf{M}_i\ddot{\mathbf{q}}_i + \mathbf{D}_i\dot{\mathbf{q}}_i + \mathbf{C}_i\mathbf{q}_i = -\mathbf{F}_{si} + \mathbf{F}_{ai}, \qquad i = x, y, z \tag{3.1}$$

where $\mathbf{q}_i$ is the displacement vector of isolated body, $\mathbf{M}_i$, $\mathbf{D}_i$, $\mathbf{C}_i$ are the inertia, damping and stiffness matrices, respectively, $\mathbf{F}_{si}$ and $\mathbf{F}_{ai}$ are the vectors of exciting and active forces describing the non-linear vibration isolator.

The n-element vector $\mathbf{q}_i$ represents the movement of elements contained in the bio-mechanical model of a human body:

$$\mathbf{q}_i = [q_{1i}, q_{2i}, ..., q_{ni}]^T, \qquad i = x, y, z \tag{3.2}$$

The n-element vectors of exciting $\mathbf{F}_{si}$ and active $\mathbf{F}_{ai}$ forces are given by the following expressions:

$$\mathbf{F}_{si} = [F_{si}, 0, ..., 0]^T, \qquad i = x, y, z \tag{3.3}$$

$$\mathbf{F}_{ai} = [F_{ai}, 0, ..., 0]^T, \qquad i = x, y, z \tag{3.4}$$

The particular non-linear exciting forces F_{si} can be described in a general form as follows:

$$F_{si} = \sum_{j=1}^{k} F_{dij}(\dot{q}_{1i} - \dot{q}_{si}) + \sum_{j=1}^{k} F_{cij}(q_{1i} - q_{si}), \qquad i = x, y, z \tag{3.5}$$

where $F_{cij}(q_{1i} - q_{si})$ are the forces of the conservative elements, and $F_{dij}(\dot{q}_{1i} - \dot{q}_{si})$ are the forces of the dissipative elements. The input vibrations are modelled as displacement q_{si} and velocity $\dot{q}_{si}$ that are generated for a specific axis of the vibration transmission x, y, z.

The individual applied forces F_{aij} are expressed as non-linear functions of the additional input signal u_i supplied from the control system:

$$F_{ai} = \sum_{j=1}^{k} F_{aij}(u_i, q_{1i} - q_{si}, \dot{q}_{1i} - \dot{q}_{si}), \qquad i = x, y, z \tag{3.6}$$

Many models of the human body are discussed in the existing literature [2–4]. Typically, they are multi-degree of freedom models developed for sitting and standing positions of the human body. This book presents a generalisation of the well-known human body models which enables to exploit a variety of bio-mechanical structures in a specific vibro-isolation system. The recommended structure of such a model is shown in Fig. 3.2.

The diagonal matrix $\mathbf{M}_i$ of size $n \times n$ contains individual masses of the human body model and is written in the following form:

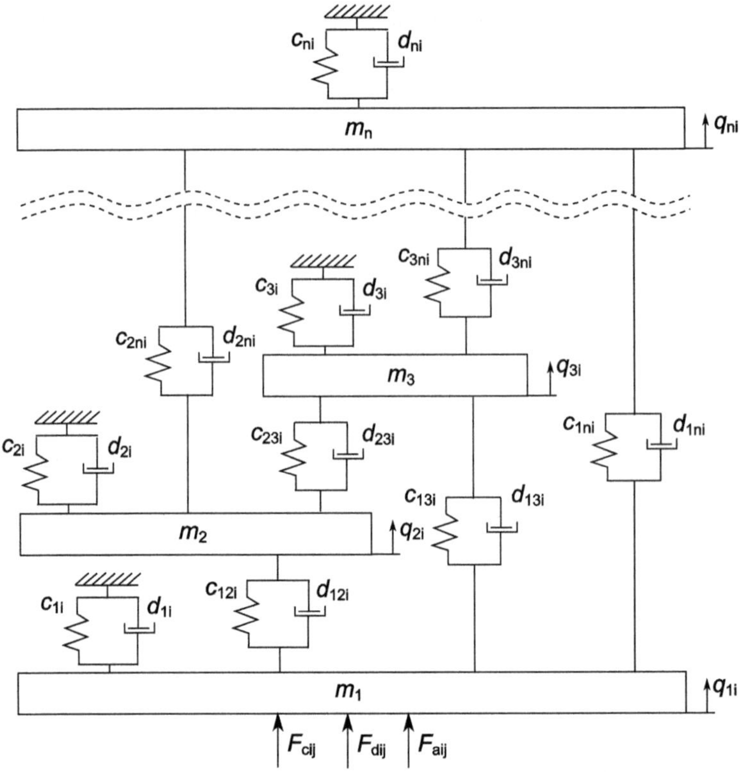

Fig. 3.2 Generalised model of a human body for one direction of the vibration transmission (i = x, y, z)

$$\mathbf{M}_\mathrm{i} = \begin{bmatrix} m_1 & 0 & 0 & \cdots & 0 \\ 0 & m_2 & 0 & \cdots & 0 \\ 0 & 0 & m_3 & \cdots & 0 \\ \cdots & \cdots & \cdots & \ddots & \cdots \\ 0 & 0 & 0 & \ldots & m_\mathrm{n} \end{bmatrix} \tag{3.7}$$

The symmetric damping matrix $\mathbf{D}_\mathrm{i}$ (size $n \times n$) shall be determined for a specific axis of the vibration transmission (i = x, y, z) as follows:

$$\mathbf{D}_\mathrm{i} = \begin{bmatrix} d_{11\mathrm{i}} & -d_{12\mathrm{i}} & -d_{13\mathrm{i}} & \cdots & -d_{1\mathrm{ni}} \\ -d_{12\mathrm{i}} & d_{22\mathrm{i}} & -d_{23\mathrm{i}} & \cdots & -d_{2\mathrm{ni}} \\ -d_{13\mathrm{i}} & -d_{23\mathrm{i}} & d_{33\mathrm{i}} & \cdots & -d_{3\mathrm{ni}} \\ \cdots & \cdots & \cdots & \ddots & \cdots \\ -d_{1\mathrm{ni}} & -d_{2\mathrm{ni}} & -d_{3\mathrm{ni}} & \ldots & d_{\mathrm{nni}} \end{bmatrix}, \quad \mathrm{i} = x, y, z \tag{3.8}$$

where the selected damping coefficients appearing on the principal diagonal should be calculated as:

$$\begin{aligned} d_{11\mathrm{i}} &= d_{1\mathrm{i}} + d_{12\mathrm{i}} + d_{13\mathrm{i}} + \cdots + d_{1\mathrm{ni}} \\ d_{22\mathrm{i}} &= d_{2\mathrm{i}} + d_{12\mathrm{i}} + d_{23\mathrm{i}} + \cdots + d_{2\mathrm{ni}} \\ d_{33\mathrm{i}} &= d_{3\mathrm{i}} + d_{13\mathrm{i}} + d_{23\mathrm{i}} + \cdots + d_{3\mathrm{ni}} \qquad \mathrm{i} = x, y, z \\ &\dots \\ d_{\mathrm{nni}} &= d_{\mathrm{ni}} + d_{1\mathrm{ni}} + d_{2\mathrm{ni}} + \cdots + d_{(\mathrm{n}-1)\mathrm{ni}} \end{aligned} \tag{3.9}$$

In succession, the stiffness matrix $\mathbf{C}_\mathrm{i}$ of size $n \times n$ have to be defined in a similar form:

$$\mathbf{C}_\mathrm{i} = \begin{bmatrix} c_{11\mathrm{i}} & -c_{12\mathrm{i}} & -c_{13\mathrm{i}} & \cdots & -c_{1\mathrm{ni}} \\ -c_{12\mathrm{i}} & c_{22\mathrm{i}} & -c_{23\mathrm{i}} & \cdots & -c_{2\mathrm{ni}} \\ -c_{13\mathrm{i}} & -c_{23\mathrm{i}} & c_{33\mathrm{i}} & \cdots & -c_{3\mathrm{ni}} \\ \dots & \dots & \dots & \ddots & \dots \\ -c_{1\mathrm{ni}} & -c_{2\mathrm{ni}} & -c_{3\mathrm{ni}} & \dots & c_{\mathrm{nni}} \end{bmatrix}, \quad \mathrm{i} = x, y, z \tag{3.10}$$

where the selected stiffness coefficients appearing on the principal diagonal must be calculated as follows:

$$\begin{aligned} c_{11\mathrm{i}} &= c_{1\mathrm{i}} + c_{12\mathrm{i}} + c_{13\mathrm{i}} + \cdots + c_{1\mathrm{ni}} \\ c_{22\mathrm{i}} &= c_{2\mathrm{i}} + c_{12\mathrm{i}} + c_{23\mathrm{i}} + \cdots + c_{2\mathrm{ni}} \\ c_{33\mathrm{i}} &= c_{3\mathrm{i}} + c_{13\mathrm{i}} + c_{23\mathrm{i}} + \cdots + c_{3\mathrm{ni}} \qquad \mathrm{i} = x, y, z \\ &\dots \\ c_{\mathrm{nni}} &= c_{\mathrm{ni}} + c_{1\mathrm{ni}} + c_{2\mathrm{ni}} + \cdots + c_{(\mathrm{n}-1)\mathrm{ni}} \end{aligned} \tag{3.11}$$

By using such a model (Eqs. 3.1–3.11), it is possible to represent the dynamic response of various vibro-isolation systems. Nonetheless, the fundamental force characteristics (F_{cij}, F_{dij}, F_{aij}) shall be determined for specific elements (e.g. hydraulic, pneumatic, electromagnetic) used to construct the suspension system.

3.4 Non-linear Models of the Typical Suspension System Components

3.4.1 *Mechanical Spring Model*

Helical springs are often used to reduce the transmission of vibration into the suspended body. The non-linear behaviour of helical springs can be obtained when the number of active coils decreases or increases with varying compression. The concept of varying the mean spring diameter in axial direction (Fig. 3.3a) is employed in order to achieve the non-constant stiffness as a function of the spring deflection.

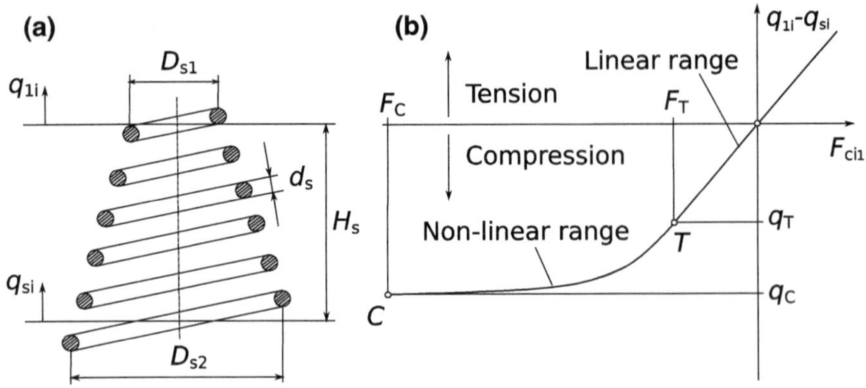

Fig. 3.3 Parameters of the conical spring (**a**) and its force–deflection characteristics (**b**)

The variable diameter of the conical spring is described as a function of the coil number n_D:

$$D(n_D) = D_{s1} + \frac{(D_{s2} - D_{s1})n_D}{n_s} \tag{3.12}$$

where D_{s1} is the mean diameter of the smallest active coil, D_{s2} is the mean diameter of the largest active coil and n_s is the overall number of active coils. The total axial deflection in the linear range (Fig. 3.3b) can be defined as follows [5]:

$$(q_{1i} - q_{si})_{lr} = \int_0^{n_s} \delta_f(n_D) dn_D \tag{3.13}$$

where $\delta_f(n_D)$ is the elementary axial deflection of the active coils. Assuming pure torsion of the spring coils, the elementary axial deflection of the active coils can be described as in the case of the cylindrical spring; however, the constant spring diameter is replaced with the variable diameter $D(n_D)$ of the conical spring [5]:

$$\delta_f(n_D) = \frac{8F_{ci1}\,(D(n_D))^3}{G_s d_s^4} \tag{3.14}$$

where F_{ci1} is the axial spring force, G_s is the Young's modulus of material elasticity and d_s is the diameter of spring wire. So the total axial deflection in the linear range reads:

$$(q_{1i} - q_{si})_{lr} = \frac{2F_{ci1}n_s}{G_s d_s^4}(D_{s1}^2 + D_{s2}^2)(D_{s1} + D_{s2}) \tag{3.15}$$

In the non-linear range (Fig. 3.3b), the active coils are gradually compressed and the total conical spring deflection is defined as [5]:

$$(q_{1i} - q_{si})_{nr} = q_f + q_s = \int_0^{n_f} \delta_f(n_D) dn_D + \int_{n_f}^{n_s} \delta_s dn_D \tag{3.16}$$

where q_f is the total axial deflection of the free coils (n_f) and q_s is the total axial deflections of the solid coils ($n_s - n_f$). The elementary axial deflection of the free coils $\delta_f(n_D)$ is already defined by Eq. (3.14). The elementary deflection of the solid coils δ_s can be calculated on the basis of the maximum geometrical spring deflection as follows:

$$\delta_s = \frac{H_s}{n_s} \tag{3.17}$$

where H_s is the initial spring height. The total conical spring deflection in the non-linear range reads:

$$(q_{1i} - q_{si})_{nr} = \frac{2F_{ci1} D_{s1}^4 n_s}{G_s d_s^4 (D_{s2} - D_{s1})} \left(\left(1 + \left(\frac{D_{s2}}{D_{s1}} - 1 \right) \frac{n_f}{n_s} \right)^4 - 1 \right) + H_s \left(1 - \frac{n_f}{n_s} \right) \tag{3.18}$$

where the number of free coils n_f is a continuous variable running from n_f to 0 at compression. If every single element reaches its maximum deflection at $n_D = n_f$, then the elementary axial deflections of the free and solid coils are assumed to be equal $\delta_f(n_f) = \delta_s$. According to Eqs. (3.14) and (3.17), the variable number of free coils can be calculated as:

$$n_f = \frac{n_s}{D_{s2} - D_{s1}} \left(\sqrt[3]{\frac{H_s G_s d_s^4}{8 F_{ci1} n_s}} - D_{s1} \right) \tag{3.19}$$

The above equations (Eqs. 3.12–3.19) have been exhaustively discussed in the paper [6]. However, in order to determine the force F_{ci1} associated with a given deflection, it is necessary to define first whether the spring displacement belongs to the linear range $(q_{1i} - q_{si})_{lr}$ or to the non-linear range $(q_{1i} - q_{si})_{nr}$.

Defining the spring forces F_T and F_C that correspond to the transition point T (the largest active coil reaches its maximum deflection) and the maximum point C (the smallest active coil reaches its maximum deflection), their values are determined using the following relations:

$$F_T = \frac{G_s d_s^4 H_s}{8 D_{s2}^3 n_s} \qquad F_C = \frac{G_s d_s^4 H_s}{8 D_{s1}^3 n_s} \tag{3.20}$$

Substituting the forces F_T and F_C into Eqs. (3.15) and (3.18), the spring displacement at the transition point T and the maximum point C can be calculated as follows:

$$q_{\mathrm{T}} = \frac{H_{\mathrm{s}}}{4D_{\mathrm{s2}}^3}(D_{\mathrm{s1}}^2 + D_{\mathrm{s2}}^2)(D_{\mathrm{s1}} + D_{\mathrm{s2}}) \tag{3.21}$$

$$q_{\mathrm{C}} = \frac{H_{\mathrm{s}} D_{\mathrm{s1}}}{4(D_{\mathrm{s2}} - D_{\mathrm{s1}})}\left(\left(1 + \left(\frac{D_{\mathrm{s2}}}{D_{\mathrm{s1}}} - 1\right)\frac{n_{\mathrm{f}}}{n_{\mathrm{s}}}\right)^4 - 1\right) + H_{\mathrm{s}}\left(1 - \frac{n_{\mathrm{f}}}{n_{\mathrm{s}}}\right) \tag{3.22}$$

Finally, the spring force is determined as a function of its deflection as follows:

$$F_{\mathrm{ci1}} = \begin{cases} \dfrac{G_{\mathrm{s}} d_{\mathrm{s}}^4 (q_{1\mathrm{i}} - q_{\mathrm{si}})}{2n_{\mathrm{s}}(D_{\mathrm{s1}}^2 + D_{\mathrm{s2}}^2)(D_{\mathrm{s1}} + D_{\mathrm{s2}})} & \text{for} \quad q_{1\mathrm{i}} - q_{\mathrm{si}} > q_{\mathrm{T}} \\ \dfrac{G_{\mathrm{s}} d_{\mathrm{s}}^4 (D_{\mathrm{s2}} - D_{\mathrm{s1}})\left(q_{1\mathrm{i}} - q_{\mathrm{si}} - H_{\mathrm{s}}\left(1 - \frac{n_{\mathrm{f}}}{n_{\mathrm{s}}}\right)\right)}{2D_{\mathrm{s1}}^4 n_{\mathrm{s}}\left(\left(1 + \left(\frac{D_{\mathrm{s2}}}{D_{\mathrm{s1}}} - 1\right)\frac{n_{\mathrm{f}}}{n_{\mathrm{s}}}\right)^4 - 1\right)} & \text{for} \quad q_{\mathrm{C}} < q_{1\mathrm{i}} - q_{\mathrm{si}} \leq q_{\mathrm{T}} \end{cases} \tag{3.23}$$

where the number of free coils n_{f} is the delayed input by given amount of time. The best accuracy is achieved when the delay is equal to the simulation step size. The resulting characteristics of the conical spring at different variations of the model parameters are shown in Fig. 3.4.

3.4.2 End-Stop Buffer Model

In order to avoid the acceleration during impact, the end-stop buffers are often mounted at the extreme points of a suspension travel [7]. These are usually

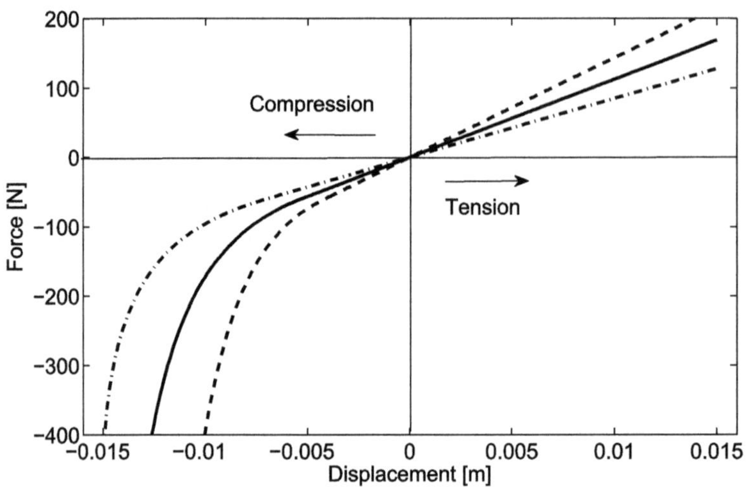

Fig. 3.4 Force characteristics of the conical spring at different variations of the model parameters: low stiffness (-·-·-), medium stiffness (—) and high stiffness (----)

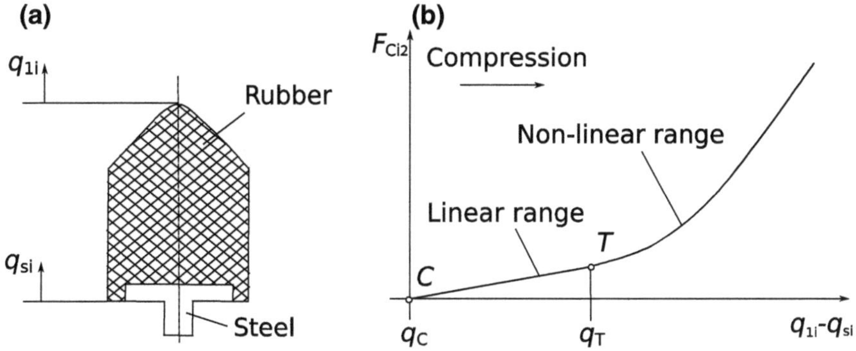

Fig. 3.5 Rubber end-stop buffer (**a**) and its force–deflection characteristics (**b**)

rubber elements vulcanised to a steel plate with a central thread (Fig. 3.5a). The rubber buffers are manufactured with different densities and inserted into the suspension mechanism to restrict a movement of the isolated body within the operating range.

The forces of end-stop buffers are expressed by using equivalent linear and non-linear springs [8] acting within certain ranges of the suspension stroke (Fig. 3.5b). A soft spring produces the reaction force F_{lr} when the suspension exceeds its free travel (point C) as follows:

$$F_{lr} = c_{b1}\,(q_{1i} - q_{si} - q_C) \tag{3.24}$$

The hard spring is modelled using a cubic non-linear stiffness, but its force F_{nr} begins to act when the suspension travel reaches the transition point T as follows:

$$F_{nr} = c_{b3}\,(q_{1i} - q_{si} - q_T)^3 + c_{b1}\,(q_{1i} - q_{si} - q_C) \tag{3.25}$$

where c_{b1} is the stiffness coefficient of linear spring, c_{b3} is the stiffness coefficient of non-linear spring, q_C is the suspension-free travel and q_T is the suspension stroke at the transition point.

Finally, the end-stop buffer force is described in the following form:

$$F_{ci2} = \operatorname{sgn}(q_{1i} - q_{si}) \begin{cases} 0 & \text{for } |q_{1i} - q_{si}| \le q_C \\ c_{b1}\,(q_{1i} - q_{si} - q_C) & \text{for } q_C < |q_{1i} - q_{si}| \le q_T \\ c_{b3}\,(q_{1i} - q_{si} - q_T)^3 + c_{b1}\,(q_{1i} - q_{si} - q_C) & \\ \qquad\qquad \text{for } q_T < |q_{1i} - q_{si}| & \end{cases} \tag{3.26}$$

The resulting force characteristics of the end-stop buffer at different variations of the model parameters are shown in Fig. 3.6.

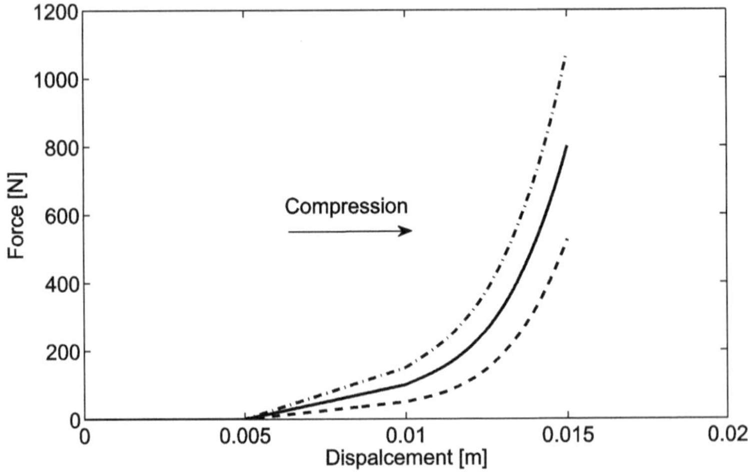

Fig. 3.6 Force characteristics of the end-stop buffer at different variations of the model parameters: low stiffness (----), medium stiffness (——) and high stiffness (-·-·-)

3.4.3 *Hydraulic Damper Model*

Hydraulic dampers are often used to dissipate energy, and therefore, these passive elements can mitigate a part of the resonant vibration. The process of energy dissipation is realised by the pressure loss across the flow restricting elements. The pressure difference between two working chambers causes the force for which the damper resists the piston movement [9]. The internal configuration of the simplified damper is shown in Fig. 3.7.

Fig. 3.7 Internal damper configuration

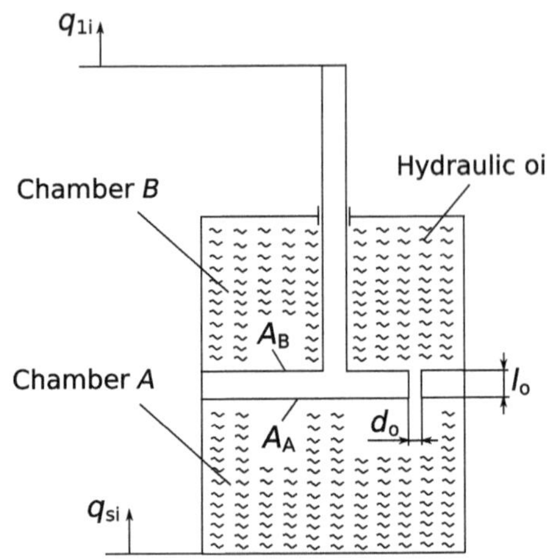

The equation describing the viscous damping force is related to the pressure loss across a simple throttle valve with a constant flow geometry [10, 11]. The pressure loss is connected with the volume flows $A_A(\dot{q}_{1i} - \dot{q}_{si})$ and $A_B(\dot{q}_{1i} - \dot{q}_{si})$ in the following way:

$$F_{di1} = \begin{cases} \zeta_{(B\to A)} \frac{\rho_o}{2} \left(\frac{A_B(\dot{q}_{1i} - \dot{q}_{si})}{A_o} \right)^2 A_B & \text{for} \quad \dot{q}_{1i} - \dot{q}_{si} > 0 \\ -\zeta_{(A\to B)} \frac{\rho_o}{2} \left(\frac{A_A(\dot{q}_{1i} - \dot{q}_{si})}{A_o} \right)^2 A_A & \text{for} \quad \dot{q}_{1i} - \dot{q}_{si} < 0 \end{cases} \tag{3.27}$$

where ρ_o is the density of liquid, $\zeta_{(B\to A)}$ and $\zeta_{(A\to B)}$ are the pressure loss coefficients between chambers A and B that are defined for a specific direction of the flow, A_A and A_B are the piston surfaces at the bottom chamber and at the piston rod-side chamber, respectively, A_o is the cross-sectional area of the orifice which is illustrated in Fig. 3.7.

The $\zeta(\text{Re})$ characteristic is employed in order to represent laminar and turbulent losses in the flow. Such a flow characteristic has to be designated for a specific geometry of the orifice. If the circular orifice is considered (at the cross-sectional area $A_o = \pi d_o^2/4$), then the pressure loss coefficient can be calculated on the basis of the Reynolds number Re [12]:

$$\zeta_{(B\to A)} = \begin{cases} \frac{64}{\text{Re}_{(B\to A)}} \left(\alpha_o + \frac{l_o}{d_o} \right) & \text{for} \quad \frac{64}{\text{Re}_{(B\to A)}} \left(\alpha_o + \frac{l_o}{d_o} \right) > 1.8 \\ 1.8 & \text{for} \quad \frac{64}{\text{Re}_{(B\to A)}} \left(\alpha_o + \frac{l_o}{d_o} \right) \le 1.8 \end{cases} \tag{3.28}$$

$$\zeta_{(A\to B)} = \begin{cases} \frac{64}{\text{Re}_{(A\to B)}} \left(\alpha_o + \frac{l_o}{d_o} \right) & \text{for} \quad \frac{64}{\text{Re}_{(A\to B)}} \left(\alpha_o + \frac{l_o}{d_o} \right) > 1.8 \\ 1.8 & \text{for} \quad \frac{64}{\text{Re}_{(A\to B)}} \left(\alpha_o + \frac{l_o}{d_o} \right) \le 1.8 \end{cases} \tag{3.29}$$

where α_o is the flow coefficient, l_o and d_o are the length and diameter of the circular orifice, respectively. The pressure loss coefficient ζ is assumed to be constant value for turbulent losses, and its value is 1.8.

The Reynolds number Re indicates the ratio between inertial and frictional forces in the flow and is defined as follows [13]:

$$\text{Re}_{(B\to A)} = \frac{4|A_B(\dot{q}_{1i} - \dot{q}_{si})|}{d_o \pi \nu_o} \qquad \text{Re}_{(A\to B)} = \frac{4|A_A(\dot{q}_{1i} - \dot{q}_{si})|}{d_o \pi \nu_o} \tag{3.30}$$

where ν_o is the kinetic viscosity of liquid. This value is also assumed as a constant. The resulting characteristics of the damper force at different variations of the model parameters are shown in Fig. 3.8.

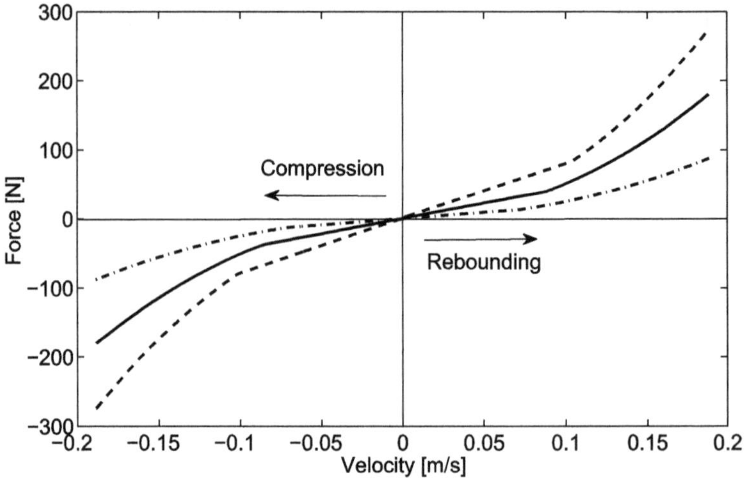

Fig. 3.8 Force characteristics of the hydraulic damper at different variations of the model parameters: low damping (-·-·-), medium damping (—) and high damping (- - - -)

3.4.4 Friction Model

Suspension mechanisms are often constructed using a linkage assembly movable through a predetermined path of travel. Friction accompanying the suspension stroke generates the force resisting its relative motion over the free travel region between the end-stop buffers. Since the system is subjected to oscillating motion, the kinetic friction should be considered as a dominant one.

The discontinuous arrangement describing the Coulomb friction model is represented by using the following function [14]:

$$F_{\mathrm{di2}} = \begin{cases} F_{\mathrm{fa}} & \text{if } \ |\dot{q}_{1\mathrm{i}} - \dot{q}_{\mathrm{si}}| \le v_{\mathrm{s}} \ \wedge \ |F_{\mathrm{fa}}| \le F_{\mathrm{fs}} \\ F_{\mathrm{fs}} \mathrm{sgn}\left(\dot{q}_{1\mathrm{i}} - \dot{q}_{\mathrm{si}}\right) & \text{if } \ |\dot{q}_{1\mathrm{i}} - \dot{q}_{\mathrm{si}}| \le v_{\mathrm{s}} \ \wedge \ |F_{\mathrm{fa}}| > F_{\mathrm{fs}} \\ F_{\mathrm{fk}} & \text{if } \ |\dot{q}_{1\mathrm{i}} - \dot{q}_{\mathrm{si}}| > v_{\mathrm{s}} \end{cases} \tag{3.31}$$

where F_{fa} is the sum of currently applied forces without friction, F_{fs} is the static friction force, F_{fk} is the kinetic friction force and v_{s} is the suspension velocity causing stick-slip instabilities. The kinetic friction force F_{fk} is further expanded considering hysteresis as the field-dependent function [8]:

$$\dot{F}_{\mathrm{fk}} = -k_{\mathrm{ff}} \left(\dot{q}_{1\mathrm{i}} - \dot{q}_{\mathrm{si}}\right) + \gamma_{\mathrm{f}} \left|\dot{q}_{1\mathrm{i}} - \dot{q}_{\mathrm{si}}\right| F_{\mathrm{fk}} + \beta_{\mathrm{f}} \left(\dot{q}_{1\mathrm{i}} - \dot{q}_{\mathrm{si}}\right) \left|F_{\mathrm{fk}}\right| \tag{3.32}$$

where k_{f} is the suspension stiffness that is responsible for the hysteresis height, γ_{f} and β_{f} are the parameters that influence the shape of hysteresis.

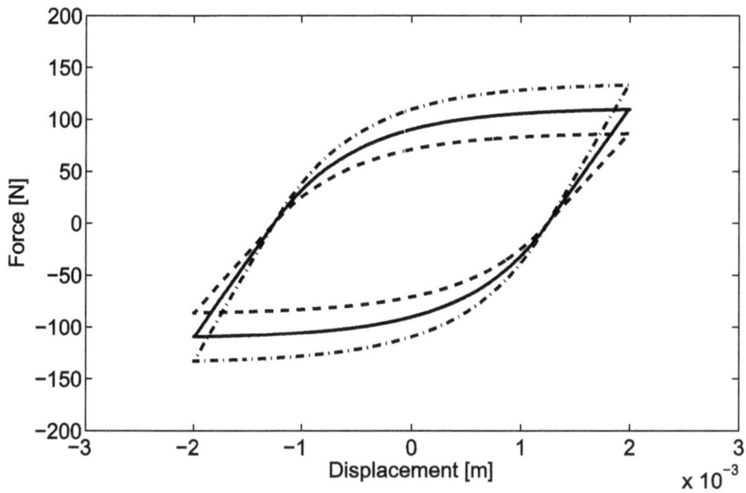

Fig. 3.9 Kinetic friction force of the suspension system at different variations of the model parameters: low friction (----), medium friction (—) and high friction (-·-·-)

The suspension stiffness k_f is defined as a function of the mass load in the following form [15]:

$$k_f = a_f \sum_{i=1}^{n} m_i + b_f \tag{3.33}$$

where m_i are the particular masses included in the human body model, a_f and b_f are the approximation coefficients. The resulting kinetic friction force at different variations of the model parameters is shown in Fig. 3.9.

3.4.5 MR Damper Model

The MR damper is an controllable shock-absorber willingly used for vibration reduction in automotive suppression applications. Such a damper is filled with magneto-rheological fluid that is controlled by an external magnetic field. The damping force can be changed very fast by varying the electric current that flows through the coil windings. In most applications [16], the solenoids are placed around the orifice between chambers A and B of the damper (Fig. 3.10). A controllable current changes the properties of a fluid flow and, in consequence, the force of the damper is regulated.

In order to model the hysteretic force–velocity characteristics of the MR damper, a model proposed in the paper [17] is employed. Such a model contains the viscous damping, spring stiffness and a hysteretic component as follows:

$$F_{ai1} = d_{mr}\left(\dot{q}_{1i} - \dot{q}_{si}\right) + c_{mr}\left(q_{1i} - q_{si}\right) + \alpha_{mr} z_{mr} + f_{mr} \tag{3.34}$$

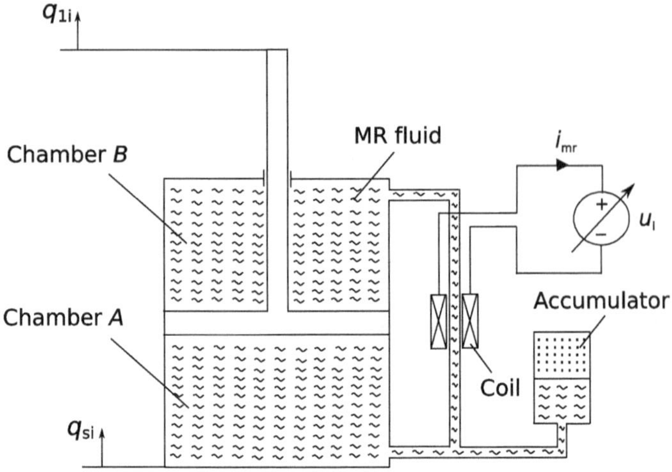

Fig. 3.10 MR damper structure

where F_{ai1} is the MR damper force, $q_{1i} - q_{si}$ and $\dot{q}_{1i} - \dot{q}_{si}$ are the relative displacement and velocity of the damper, respectively, d_{mr} is the viscous damping coefficient, c_{mr} is the stiffness coefficient, α_{mr} is the scale factor of the hysteresis, z_{mr} is the hysteretic variable and f_{mr} is the force offset. The hysteretic variable can be given by the hyperbolic tangent function [17]:

$$z_{mr} = \tanh\left(\beta_{mr}\left(\dot{q}_{1i} - \dot{q}_{si}\right) + \delta_{mr}\text{sign}\left(q_{1i} - q_{si}\right)\right) \tag{3.35}$$

where β_{mr} and δ_{mr} are the model parameters giving the shape of hysteresis. Based on the experimental data, the least square approximation method may be employed to determine the appropriate parameters d_{mr} and α_{mr} for the analytical model [18]:

$$d_{mr} = a_1 i_{mr} + a_0 \tag{3.36}$$

$$\alpha_{mr} = b_2 i_{mr}^2 + b_1 i_{mr} + b_0 \tag{3.37}$$

where a_1, a_0 and b_2, b_1, b_0 are the polynomial coefficients expressed with respect to the input current i_{mr}. These coefficients have to be evaluated experimentally for a specific MR damper type.

The resulting force characteristics of the MR damper at different currents flowing through the coil windings are shown in Fig. 3.11.

The MR damper dynamics is described by using the first-order inertial element [13] with the time constant t_{mr}. The following function is used in order to model the system delay in response to an electrical control signal u_i:

$$t_{mr}\dot{i}_{mr} + i_{mr} = k_{mr}u_i \tag{3.38}$$

where k_{mr} is the static gain of the MR damper.

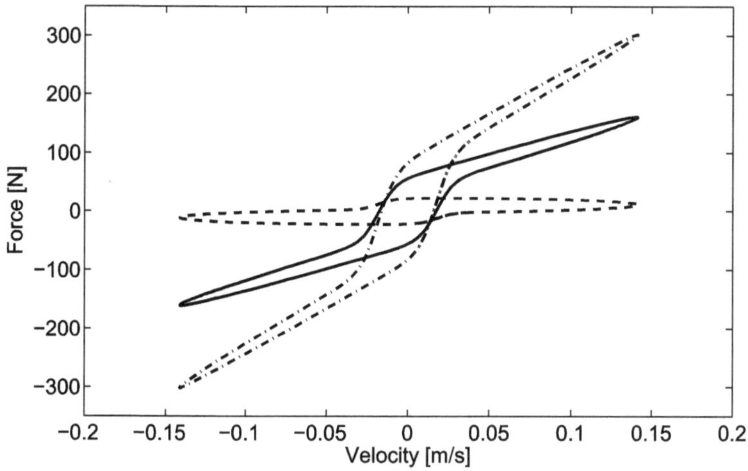

Fig. 3.11 Force characteristics of the MR damper at different currents flowing through the coil windings: 0 A (----) , 2 A (—) and 4 A (-·-·-)

3.4.6 Pneumatic Muscle Model

Pneumatic muscles are the flexible actuators that are utilised more and more in mechatronic engineering. These actuators contain the retractile tubing that is manufactured by using a rubber material with specially arranged fibres. If the air-pressure is activated, then a rubber material is expanded in the radial direction and a tension force is generated axially along the muscle. This action mimics the human muscle, and the highest tension force is accessible at its nominal length before contraction. The gradual contraction of pneumatic muscle causes decreasing of its force. Parameters of the pneumatic muscle at the nominal length and after contraction are shown in Fig. 3.12.

The contraction force generated in the muscle can be calculated on the basis of its virtual work [19]:

$$dW_p + dW_f = 0 \tag{3.39}$$

where dW_p is the elementary work driven by compressed air when the muscle volume is changed, dW_f is the elementary work provided by external force during contraction of the pneumatic muscle. Thus, the following relation may be expressed [19]:

$$(p_m - p_0)dV_m + F_{ai2}d(q_{1i} - q_{si}) = 0 \tag{3.40}$$

where p_m is the absolute air-pressure in the muscle, p_0 is the atmospheric pressure, dV_m is the volume change, F_{ai2} is the contraction force and $d(q_{1i} - q_{si})$ is the

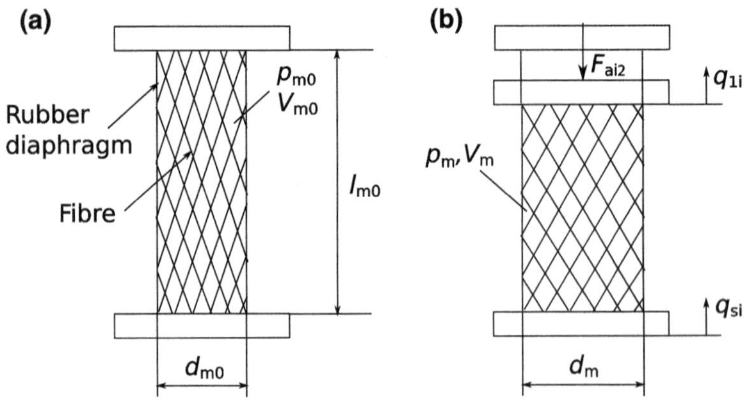

Fig. 3.12 Parameters of the pneumatic muscle at the nominal length (**a**) and after contraction (**b**)

reduction of muscle relative displacement. Rearranging the Eq. (3.40), the muscle force is determined by the following expression:

$$F_{\mathrm{ai2}} = -(p_{\mathrm{m}} - p_0)\frac{dV_{\mathrm{m}}}{d(q_{1\mathrm{i}} - q_{\mathrm{si}})} \tag{3.41}$$

The muscle volume V_{m} is calculated on the basis of its changeable diameter d_{m} and length $q_{1\mathrm{i}} - q_{\mathrm{si}}$ in such a way:

$$V_{\mathrm{m}} = \frac{\pi}{4} d_{\mathrm{m}}^2 (q_{1\mathrm{i}} - q_{\mathrm{si}}) \tag{3.42}$$

If the constant length a_{m} of the fibre is assumed (Fig. 3.13), then cylindrical volume of the pneumatic muscle may be interpreted as follows:

$$V_{\mathrm{m}} = \frac{1}{4\pi}\left(a_{\mathrm{m}}^2 - (q_{1\mathrm{i}} - q_{\mathrm{si}})^2\right)(q_{1\mathrm{i}} - q_{\mathrm{si}}) \tag{3.43}$$

Differentiating the Eq. (3.43) and substituting the trigonometric functions which are presented in Fig. 3.13, the succeeding equation is acquired:

$$\frac{dV_{\mathrm{m}}}{d(q_{1\mathrm{i}} - q_{\mathrm{si}})} = \frac{\pi}{4} d_{\mathrm{m}}^2 \left(1 - \frac{2}{\tan^2 \beta_{\mathrm{m}}}\right) \tag{3.44}$$

with the changeable diameter d_{m} and angle β_{m} that are determined by means of the muscle geometry before contraction:

$$d_{\mathrm{m}} = d_{\mathrm{m0}} \frac{\sin \beta_{\mathrm{m}}}{\sin \beta_{\mathrm{m0}}}, \qquad \beta_{\mathrm{m}} = \arccos\left(\frac{q_{1\mathrm{i}} - q_{\mathrm{si}}}{l_{\mathrm{m0}}} \cos \beta_{\mathrm{m0}}\right) \tag{3.45}$$

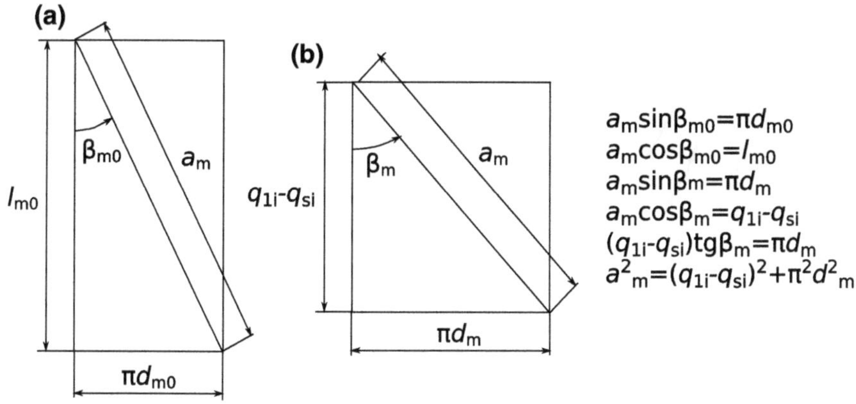

Fig. 3.13 Lateral surface of the pneumatic muscle at the nominal length (**a**) and after contraction (**b**) shown in the case of a single fibre

where d_{m0}, β_{m0} and l_{m0} are the nominal values of parameters characterises specific type of the pneumatic muscle.

Force characteristics of the pneumatic muscle can be obtained by substituting Eq. (3.44) into Eq. (3.41) in the following way:

$$F_{\mathrm{ai2}} = -\frac{\pi}{4}(p_{\mathrm{m}} - p_{\mathrm{O}})d_{\mathrm{m}}^2\left(1 - \frac{2}{\tan^2\beta_{\mathrm{m}}}\right) \tag{3.46}$$

Whereas the contracting or stretching force also depends on the geometric and material properties of the rubber and fibre [20], an extension of the model expressed by Eq. (3.46) is recommended by means of the subsequent relation:

$$f_{\mathrm{m}} = \mu_{\mathrm{m}}\left(1 - \exp\left(\frac{l_{\mathrm{m0}} - q_{1\mathrm{i}} - q_{\mathrm{si}}}{\eta_{\mathrm{m}} l_{\mathrm{m0}}}\right)\right) - \nu_{\mathrm{m}} \tag{3.47}$$

where μ_{m}, η_{m} and ν_{m} are the parameters shaping the force characteristics. Finally, the modified force of pneumatic muscle is defined as follows:

$$F_{\mathrm{ai2}} = -\frac{\pi}{4}(p_{\mathrm{m}} - p_{\mathrm{O}})d_{\mathrm{m}}^2\left(1 - \frac{2}{\tan^2\beta_{\mathrm{m}}}\right) + f_{\mathrm{m}} \tag{3.48}$$

The resulting force characteristics of the pneumatic muscle at different air-pressure are shown in Fig. 3.14.

The dynamic properties of the proportional pressure control valve are modelled as the first-order inertial element [13] with the time constant t_{m} that describes the

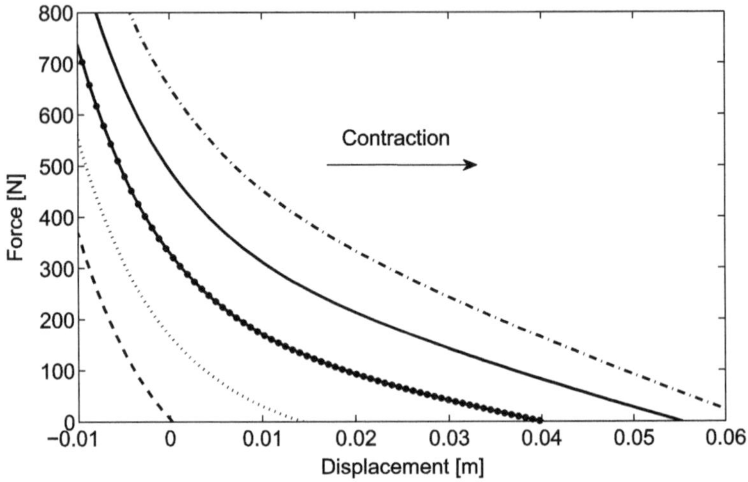

Fig. 3.14 Force characteristics of the pneumatic muscle at different air-pressure: $1 \cdot 10^5$ Pa (----), $3 \cdot 10^5$ Pa (·····), $5 \cdot 10^5$ Pa (-◆-), $7 \cdot 10^5$ Pa (—), $9 \cdot 10^5$ Pa (-·-·-)

system dynamics in response to an electrical input signal u_i. The equation describing the proportional pressure control valve is written as follows:

$$t_m \dot{p}_m + p_m = k_m u_i \tag{3.49}$$

where k_m is the static gain of the proportional pressure control valve.

3.5 Example: Modelling and Simulation of a Horizontal Seat Suspension

3.5.1 Model of the Passive Suspension

In Fig. 3.15, a physical model of an exemplary horizontal suspension containing helical springs, end-stop buffers and hydraulic shock-absorber is shown. The seat suspension model takes into account the spring force F_{cx1}, the forces from end-stop buffers F_{cx2} limiting the maximum relative displacement, the damping force of hydraulic shock-absorber F_{dx1} and the friction force of suspension mechanism F_{dx2}. In order to facilitate verification of the suspension model, the single degree of freedom body mass m_1 is used in this section.

The equation of motion for the passive suspension system, as an equilibrium of forces acting on the isolated body, reads:

$$m_1 \ddot{q}_{1x} = -2F_{cx1} - F_{cx2} - F_{dx1} - F_{dx2} \tag{3.50}$$

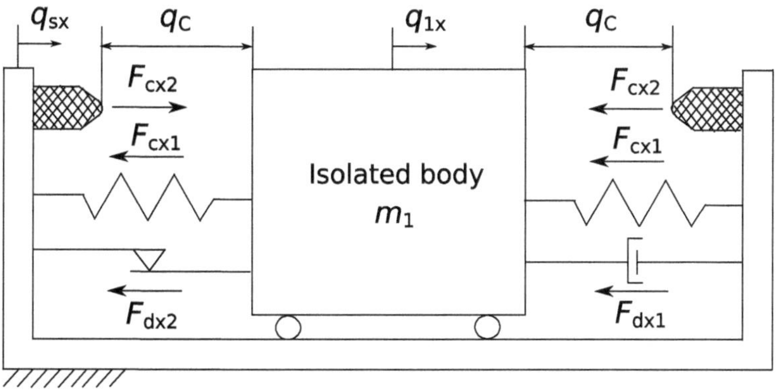

Fig. 3.15 Physical model of the horizontal seat suspension

where q_{1x} is the horizontal displacement of the isolated body. The mathematical models of the basic forces in the passive system have been presented in Sect. 3.4. The evaluation of numerical values of the model parameters is based on measurements.

3.5.2 Evaluation of the Model Parameters

An evaluation of the model parameters is determined with the help of the force–deflection measurement of the suspension system. The force sensor is affixed to the isolated body which remains immovable. The lower part of the suspension system is excited with different amplitudes and frequencies. The displacement sensor is installed for evaluating the actual suspension deflection. The experimental set-up used for the evaluation of the system characteristics is shown in Fig. 3.16.

The force characteristics of helical spring are determined by using the force–displacement measurement with triangle cycling of the suspension system at a frequency of 0.01 Hz and an amplitude of ±0.015 m. Two tension springs are installed

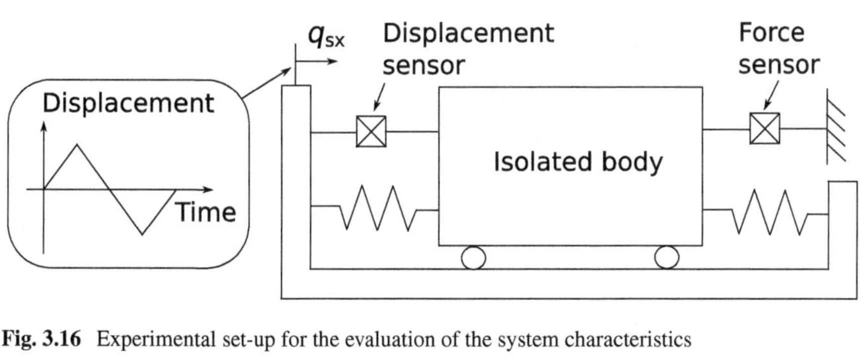

Fig. 3.16 Experimental set-up for the evaluation of the system characteristics

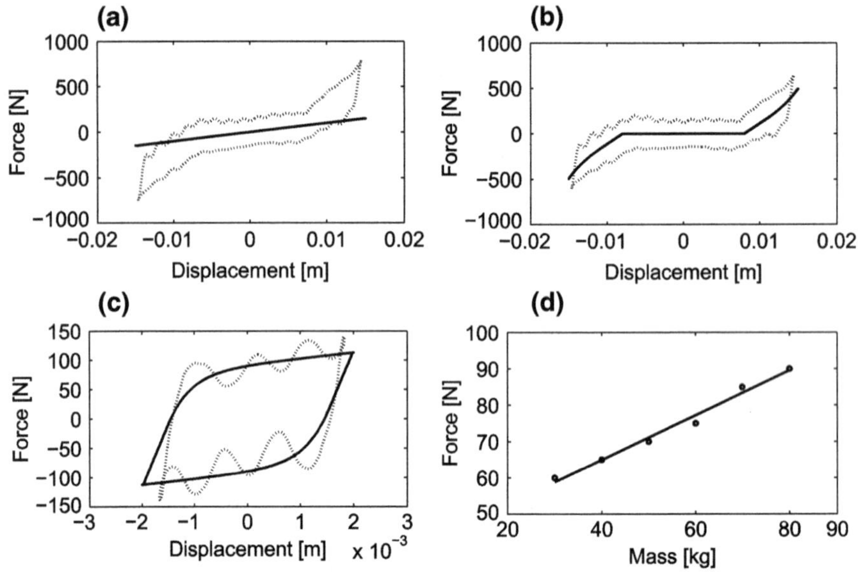

Fig. 3.17 Stiffness characteristics of the helical springs (**a**), stiffness characteristics of the end-stop buffers (**b**), friction characteristics of the suspension system (**c**) obtained using simulation model (—) and measurement (·····), measured (○) and approximated (—) friction force for different masses loaded the suspension system (**d**)

in the horizontal seat suspension; therefore, the linear part of the mechanical spring model (Eq. 3.23) is used for this exemplary system. The simulation and measurement results are shown in Fig. 3.17a.

The stiffness characteristics of the end-stop buffer are determined while the suspension system without helical springs is deflected using the same test conditions. The force and displacement of the system are measured during this laboratory investigation. The simulation and measurement results, that correspond to the end-stop buffer model described by Eq. (3.26), are presented in Fig. 3.17b.

The friction characteristics are measured at a frequency of 1 Hz and an amplitude of ± 0.002 m again with triangle cycling of the suspension mechanism. The measured hysteresis of the friction force in displacement domain (Fig. 3.17c) is fitted to the simulation results by appropriate selection of the model parameters (Eq. 3.32). The measured and approximated friction force, that are evaluated for different masses loaded the suspension system, are presented in Fig. 3.17d.

The force of shock-absorber in velocity domain is evaluated by sinusoidal cycling of the damper. The shock-absorber is moved at an amplitude of ± 0.0125 m and two different frequencies, i.e. 0.83 and 1.66 Hz. The force and velocity sensors are used for measuring the maximum damping force at compressing and rebounding of the shock-absorber (Fig. 3.18a). The modelled damping characteristics using Eqs. (3.27)–(3.30) and the measured data are shown in Fig. 3.18b.

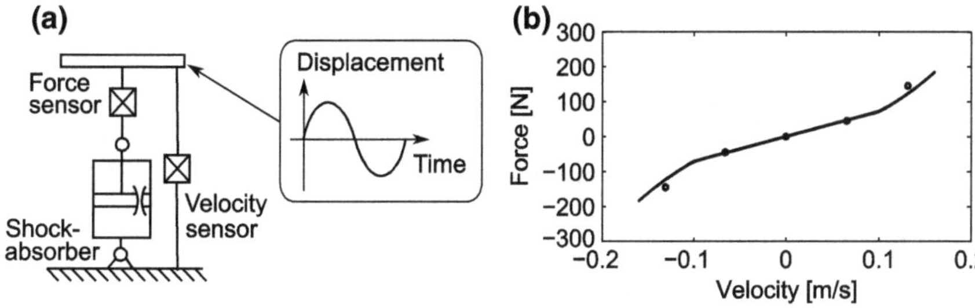

Fig. 3.18 Experimental set-up for the evaluation of the damper force (**a**), damping characteristics of the hydraulic shock-absorber (**b**) obtained using simulation model (—) and measurements (∘)

Each parameter value used by the seat suspension model is shown in Table 3.1. Moreover, all calculations presented in this section are related to a specific type of horizontal suspension, which means that all parameters have to be evaluated experimentally for a specific suspension type.

3.5.3 Model Verification

The non-linear ordinary differential equations (ODE) describing the dynamic behaviour of horizontal seat suspension system are modelled in the MATLAB-Simulink ® software package. Such a software allows to build simulation models by using the graphical interface, and the system equations are solved numerically employing Bogacki–Shampine solver [21] with the fixed-step size (step time of 1 ms).

In order to verify the functioning of the engineering solution of the seat suspension system as described above, experimental tests are carried out with the use of an experimental set-up shown in Fig. 3.19. The horizontal seat suspension is loaded with body masses 30, 55 and 80 kg.

The measurements are performed for the excitation signals that are similar to the white noise and band-limited noise. These excitation signals are obtained using a signal generator with the random waveform. Spectral characteristics of normally distributed random signals are subsequently formed by the Butterworth filters: high-pass (HP) and low-pass (LP) in the range of frequency 0.5–10 Hz. For the purpose of the model verification, three different excitation intensities are applied, i.e. WN1x, WN3x and WN5x. The power spectral densities of the excitation signals are shown in Fig. 3.20. The desired and obtained root mean square values of the acceleration signal are defined in Table 3.2.

The measured and simulated signals are: the acceleration of the vibration platform, the acceleration of the isolated body, the relative displacement of the suspension system and the absolute displacement of the vibration platform. Based on the accel-

Table 3.1 Model parameters of the horizontal seat suspension

System component	Parameter	Value	Unit
Helical spring	Mean diameter of the smallest active coil (D_{s1})	$11.9 \cdot 10^{-3}$	m
	Mean diameter of the largest active coil (D_{s2})	$11.9 \cdot 10^{-3}$	m
	Shear modulus of elasticity (G_s)	$121 \cdot 10^9$	N/m^2
	Wire diameter (d_s)	$1.8 \cdot 10^{-3}$	m
	Number of active coils (n_s)	19	–
End-stop buffer	Linear stiffness coefficient (c_{b1})	$28 \cdot 10^3$	N/m
	Cubic stiffness coefficient (c_{b3})	$4 \cdot 10^8$	N/m^3
	Suspension free travel (q_C)	$8 \cdot 10^{-3}$	m
	Suspension stroke at the transition point (q_T)	$10 \cdot 10^{-3}$	m
Hydraulic shock-absorber	Surface of shock-absorber piston, side A (A_A)	$1.77 \cdot 10^{-4}$	m^2
	Surface of shock-absorber piston, side B (A_B)	$1.64 \cdot 10^{-4}$	m^2
	Orifice diameter (d_o)	$1 \cdot 10^{-3}$	m
	Orifice length (l_o)	$13 \cdot 10^{-3}$	m
	Flow coefficient (α_o)	0.667	–
	Kinetic viscosity of a liquid (ν_o)	4.6×10^{-5}	m^2/s
	Density of liquid (ρ_o)	890	kg/m^3
Friction	Static friction force (F_{fs})	150	N
	Relative slipping velocity (v_s)	$1 \cdot 10^{-5}$	m/s
	First approximation coefficient (a_f)	$2.45 \cdot 10^3$	N/(m · kg)
	Second approximation coefficient (b_f)	$161 \cdot 10^3$	N/m
	First parameter regulating the shape of a hysteretic friction model (β_f)	$2 \cdot 10^3$	m^{-1}
	Second parameter regulating the shape of a hysteretic friction model (γ_f)	$2 \cdot 10^3$	m^{-1}

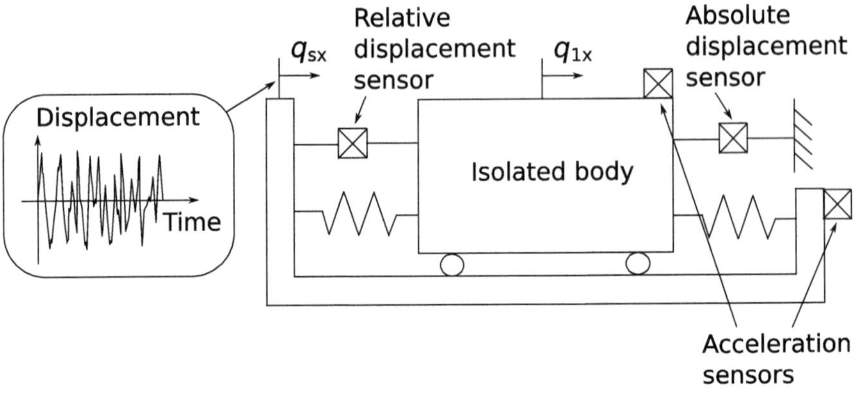

Fig. 3.19 Experimental set-up for the evaluation of the system dynamics

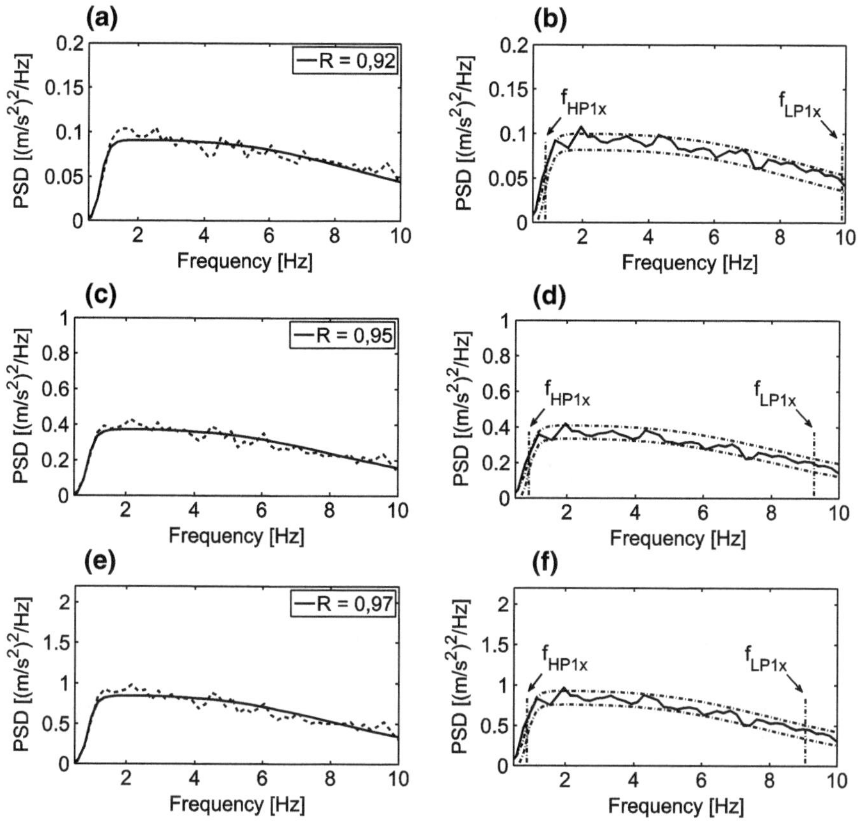

Fig. 3.20 Power spectral densities of measured (----), estimated (——) (left-hand side) and simulated (——) (right-hand side) input vibrations for the spectral classes: WN1x (**a–b**), WN3x (**c–d**), WN5x (**e–f**) and their tolerances (-·-·-)

Table 3.2 Parameters of the horizontal input vibration generated for the purpose of model verification

Input vibration	Signal generator	High-pass filter		Low-pass filter		Results	
						Desired	Obtained
	σ_{x1}^2, $\left(\frac{m}{s^2}\right)^2$	f_{HPx1}, Hz	n_{HPx1}	f_{LPx1}, Hz	n_{LPx1}	$(\ddot{q}_{sx})_{RMS}$, $\frac{m}{s^2}$	
WN1x	0.95	0.88	4	9.89	2	0.98	1.02
WN3x	3.45	0.89	4	9.26	2	1.86	1.91
WN5x	7.81	0.89	4	9.06	2	2.79	2.86

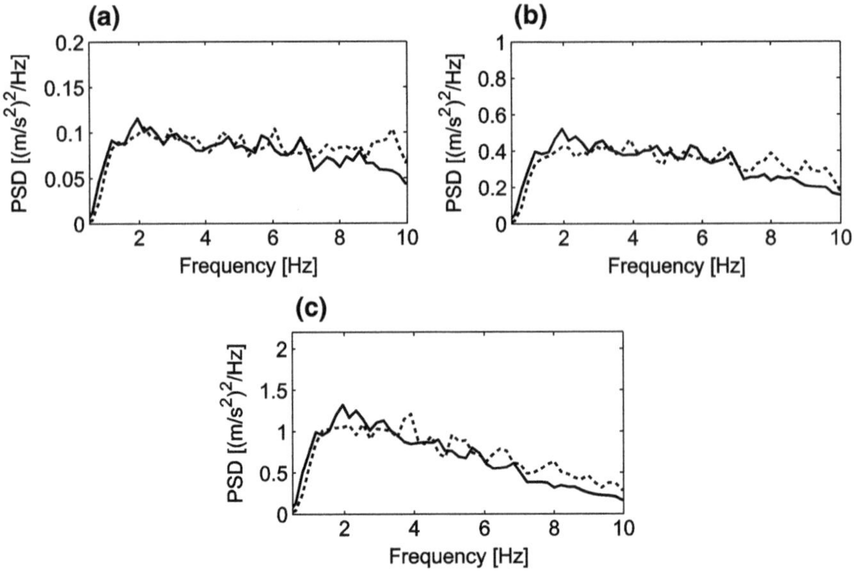

Fig. 3.21 Simulated (—) and measured (----) power spectral densities for different spectral classes: WN1x (**a**), WN3x (**b**), WN5x (**c**), mass load 30 kg

eration signals coming from the vibration platform and suspended mass, the power spectral densities are evaluated. The simulation and experimental results for selected excitation signal are shown in Figs. 3.21, 3.22 and 3.23.

3.6 Apparent Mass Model of the Seated Human Body

The apparent mass model is employed to measure the bio-dynamic response of the seated human body that is exposed to a single-axis vibration. Such a research methodology relies on measuring the force that is required to generate a given movement of the human body (Fig. 3.24). The apparent mass is usually determined by using the signals acquired from accelerometers and force transducers [22].

Then, the apparent mass can be defined as a complex ratio of the force $F_{1\mathrm{i}}(2\pi f)$ applied at specific frequency f to the obtained vibration acceleration $\ddot{q}_{1\mathrm{i}}(2\pi f)$. Both signals have to measured at the same point and in the same axis of vibration transmission as follows [23]:

$$M_{\mathrm{i}}(2\pi f) = \frac{F_{1\mathrm{i}}(2\pi f)}{\ddot{q}_{1\mathrm{i}}(2\pi f)}, \qquad \mathrm{i} = x, y, z \tag{3.51}$$

where f is the vibration frequency in Hz. The apparent mass method can be successfully utilised to indicate the resonances occurring in the human body [7]. Although

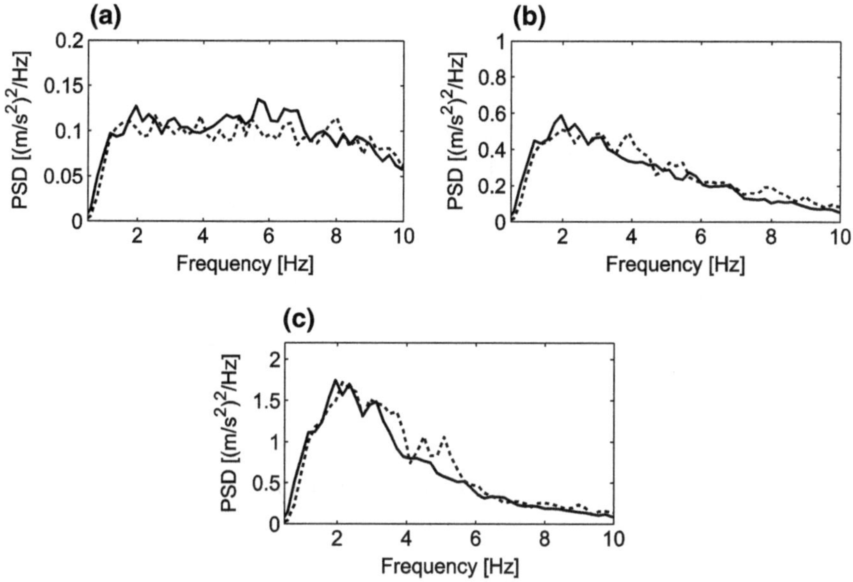

Fig. 3.22 Simulated (—) and measured (----) power spectral densities for different spectral classes: WN1x (**a**), WN3x (**b**), WN5x (**c**), mass load 55 kg

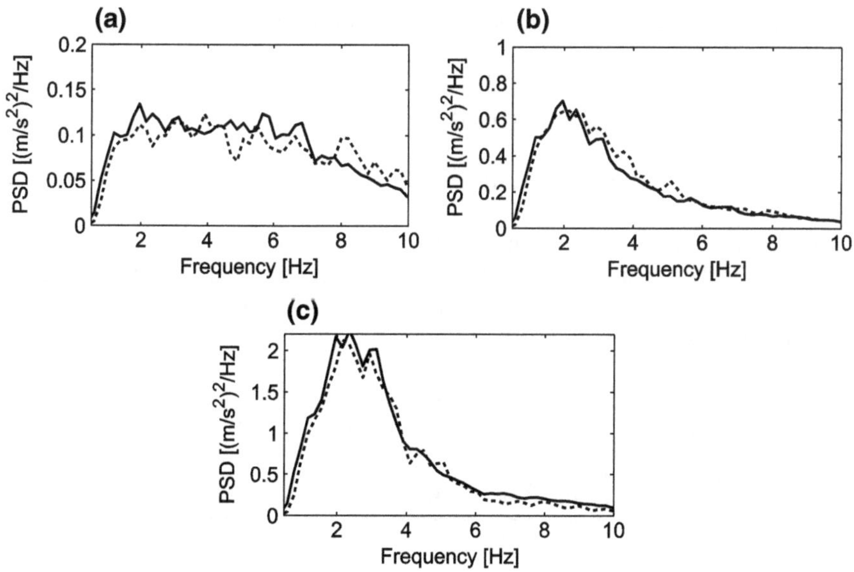

Fig. 3.23 Simulated (—) and measured (----) power spectral densities for different spectral classes: WN1x (**a**), WN3x (**b**), WN5x (**c**), mass load 80 kg

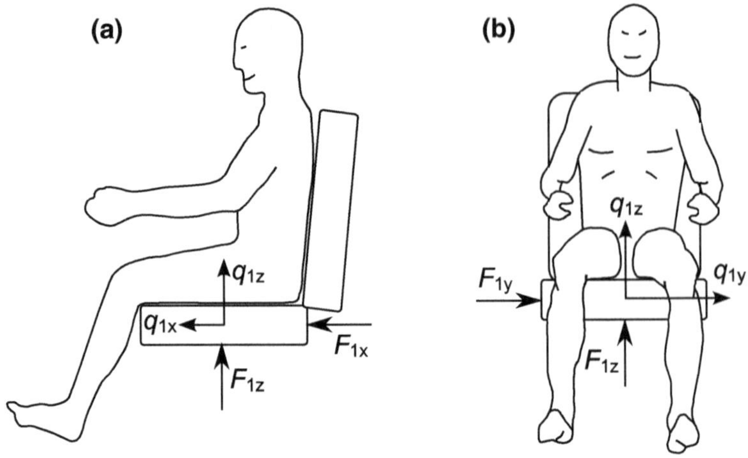

Fig. 3.24 Test person of the seated human body: side view (**a**) and front view (**b**)

the apparent mass for a rigid body is equal to its weight, it demonstrates vibration amplitudes and resonant frequencies of a flexible structure [24].

For the purpose of evaluating the apparent mass $M_i(2\pi f)$, the state-space model is engaged as a set of input, output and state variables related by first-order differential equations. The state variables are defined in the following vector form:

$$\begin{aligned} \mathbf{x}_{1i} &:= [q_{1i}, \dot{q}_{1i}]^T \\ \mathbf{x}_{2i} &:= [q_{2i}, \dot{q}_{2i}]^T \qquad\qquad i = x, y, z \\ &\dots \\ \mathbf{x}_{ni} &:= [q_{ni}, \dot{q}_{ni}]^T \end{aligned} \tag{3.52}$$

where q_{1i}, ..., q_{ni} and $\dot{q}_{1i}$, ..., $\dot{q}_{ni}$ are the particular displacements and velocities corresponding to the human body model presented in Sect. 3.3 (Fig. 3.2). Assuming that the applied force F_{1i} is an external input to the model, the state equation can be defined as follows:

$$\underbrace{\begin{bmatrix} \dot{\mathbf{x}}_{1i} \\ \dot{\mathbf{x}}_{2i} \\ \dots \\ \dot{\mathbf{x}}_{ni} \end{bmatrix}}_{\dot{\mathbf{x}}_{si}} = \underbrace{\begin{bmatrix} \mathbf{A}_{11i} & \mathbf{A}_{12i} & \cdots & \mathbf{A}_{1ni} \\ \mathbf{A}_{12i} & \mathbf{A}_{22i} & \cdots & \mathbf{A}_{2ni} \\ \dots & \dots & \dots & \dots \\ \mathbf{A}_{1ni} & \mathbf{A}_{2ni} & \dots & \mathbf{A}_{nni} \end{bmatrix}}_{\mathbf{A}_{si}} \underbrace{\begin{bmatrix} \mathbf{x}_{1i} \\ \mathbf{x}_{2i} \\ \dots \\ \mathbf{x}_{ni} \end{bmatrix}}_{\mathbf{x}_{si}} + \underbrace{\begin{bmatrix} \mathbf{B}_{1i} \\ 0 \\ \dots \\ 0 \end{bmatrix}}_{\mathbf{B}_{si}} [F_{1i}] \tag{3.53}$$

where $\mathbf{x}_{si}$ is the state vector, $\mathbf{A}_{si}$ is state (system) matrix and $\mathbf{B}_{si}$ is the input vector. The principal diagonal elements of the state matrix should be calculated by using the following relationships:

$$\mathbf{A}_{11\mathrm{i}} = \begin{bmatrix} 0 & 1 \\ -\frac{c_{11\mathrm{i}}}{m_1} & -\frac{d_{11\mathrm{i}}}{m_1} \end{bmatrix} \quad \mathbf{A}_{22\mathrm{i}} = \begin{bmatrix} 0 & 1 \\ -\frac{c_{22\mathrm{i}}}{m_2} & -\frac{d_{22\mathrm{i}}}{m_2} \end{bmatrix} \tag{3.54}$$

$$\cdots \quad \mathbf{A}_{\mathrm{nni}} = \begin{bmatrix} 0 & 1 \\ -\frac{c_{\mathrm{nni}}}{m_\mathrm{n}} & -\frac{d_{\mathrm{nni}}}{m_\mathrm{n}} \end{bmatrix}$$

Another elements of the state matrix are defined as follows:

$$\mathbf{A}_{12\mathrm{i}} = \begin{bmatrix} 0 & 0 \\ \frac{c_{12\mathrm{i}}}{m_1} & \frac{d_{12\mathrm{i}}}{m_1} \end{bmatrix} \quad \mathbf{A}_{1\mathrm{ni}} = \begin{bmatrix} 0 & 0 \\ \frac{c_{1\mathrm{ni}}}{m_1} & \frac{d_{1\mathrm{ni}}}{m_1} \end{bmatrix} \tag{3.55}$$

$$\mathbf{A}_{2\mathrm{ni}} = \begin{bmatrix} 0 & 0 \\ \frac{c_{2\mathrm{ni}}}{m_2} & \frac{d_{2\mathrm{ni}}}{m_2} \end{bmatrix} \quad \mathbf{A}_{\mathrm{nni}} = \begin{bmatrix} 0 & 0 \\ \frac{c_{\mathrm{nni}}}{m_\mathrm{n}} & \frac{d_{\mathrm{nni}}}{m_\mathrm{n}} \end{bmatrix}$$

Successively, the first element of the input vector is specified using the following component:

$$\mathbf{B}_{1\mathrm{i}} = \begin{bmatrix} 0 \\ \frac{1}{m_1} \end{bmatrix} \tag{3.56}$$

The output equation is formulated on the assumption that the vibration acceleration $y_1 := \ddot{q}_{1\mathrm{i}}$ is chosen as the output variable:

$$[y_{1\mathrm{i}}] = \underbrace{\begin{bmatrix} \mathbf{C}_{11\mathrm{i}} \; \mathbf{C}_{12\mathrm{i}} \cdots \mathbf{C}_{1\mathrm{ni}} \end{bmatrix}}_{\mathbf{C}_{\mathrm{si}}} \underbrace{\begin{bmatrix} \mathbf{x}_{1\mathrm{i}} \\ \mathbf{x}_{2\mathrm{i}} \\ \cdots \\ \mathbf{x}_{\mathrm{ni}} \end{bmatrix}}_{\mathbf{x}_{\mathrm{si}}} + \underbrace{\begin{bmatrix} \mathbf{D}_{1\mathrm{i}} \\ 0 \\ \cdots \\ 0 \end{bmatrix}}_{\mathbf{D}_{\mathrm{si}}} [F_{1\mathrm{i}}] \tag{3.57}$$

where $\mathbf{C}_{\mathrm{si}}$ is the output matrix and $\mathbf{D}_{\mathrm{si}}$ is the feedthrough (feedforward) vector. The individual elements of the output vector are calculates using following relationships:

$$\mathbf{C}_{11\mathrm{i}} = \begin{bmatrix} -\frac{c_{11\mathrm{i}}}{m_1} & -\frac{d_{11\mathrm{i}}}{m_1} \end{bmatrix} \quad \mathbf{C}_{12\mathrm{i}} = \begin{bmatrix} \frac{c_{12\mathrm{i}}}{m_1} & \frac{d_{12\mathrm{i}}}{m_1} \end{bmatrix} \tag{3.58}$$

$$\cdots \quad \mathbf{C}_{1\mathrm{ni}} = \begin{bmatrix} \frac{c_{1\mathrm{ni}}}{m_1} & \frac{d_{1\mathrm{ni}}}{m_1} \end{bmatrix}$$

The first element of the feedforward vector is described by using the following component:

$$\mathbf{D}_{1\mathrm{i}} = \begin{bmatrix} \frac{1}{m_1} \\ 0 \end{bmatrix} \tag{3.59}$$

Thus, the created state-space model describing multi-body dynamics of vibration structure (Eqs. 3.52–3.59) is converted to the corresponding transfer function:

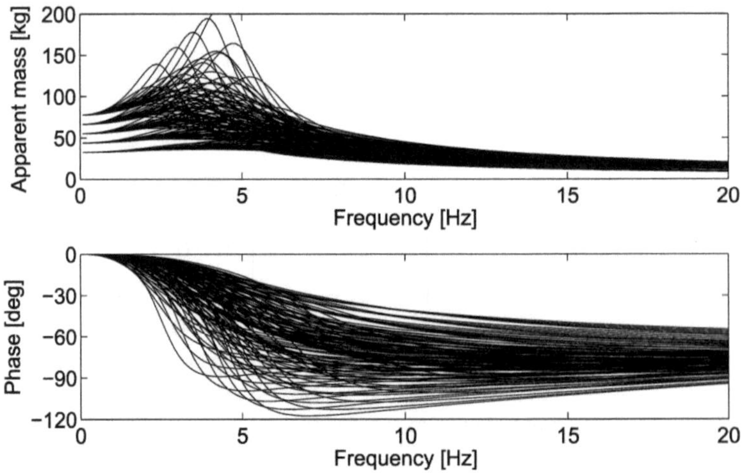

Fig. 3.25 Apparent mass and phase at different variations of the model parameters: mass ±50 %, damping ±50 %, stiffness ±50 % of their nominal values

$$H_i(s) = \frac{\ddot{q}_{1i}(s)}{F_{1i}(s)} = \mathbf{C}_{si}(s\mathbf{I} - \mathbf{A}_{si})^{-1}\mathbf{B}_{si} + \mathbf{D}_{si}, \qquad i = x, y, z \tag{3.60}$$

where s is the Laplace variable and $\mathbf{I}$ is the identity matrix. The denominator of the transfer function (Eq. 3.60) is determined as equal to the characteristic polynomial established by using the determinant of $s\mathbf{I} - \mathbf{A}_{si}$.

Ultimately, the apparent mass $M_i(2\pi f)$ and phase shift $\phi_i(2\pi f)$ representing the human body model (Fig. 3.2) are given as the following expressions:

$$M_i(2\pi f) = \left|H_i^{-1}(s)\right|, \qquad \phi_i(2\pi f) = \angle H_i^{-1}(s), \qquad i = x, y, z \tag{3.61}$$

Moduli and phases of the elaborated apparent masses model are presented in Fig. 3.25. These results are obtained for an exemplary four degrees of freedom model with different variations imposed on its numerical parameters. As clearly follows from this figure, a suitable selection of the model parameters leads to some general results in terms of frequency dependence of human sensitivity to vibration.

3.7 Example: Models of the Apparent Mass for Seated Human Body

3.7.1 Apparent Mass Model Under Vertical Vibration

The International Standard ISO 5982 [23] constitutes the human body model (Fig. 3.26) which consists of four interconnected masses by means of linear springs and dampers. The force F_{1z} is considered as a model input, and such a force is applied

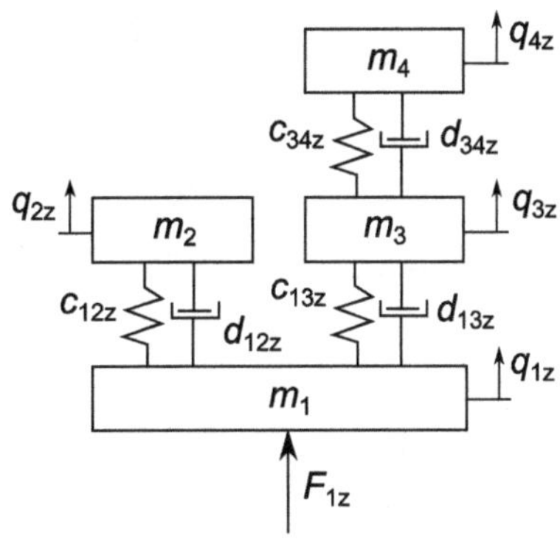

Fig. 3.26 Bio-mechanical model of the human body according to ISO 5982 [23]

Table 3.3 Numerical values of the model parameters (total body mass of 75 kg) [23]

	Mass, kg	Damping, Ns/m		Stiffness, N/m	
m_1	2	d_{12z}	1.39×10^3	c_{12z}	3.62×10^4
m_2	45 (30, 56)	d_{13z}	387	c_{13z}	9.99×10^3
m_3	6	d_{34z}	234	c_{34z}	3.44×10^4
m_4	2				

to mass m_1. The resulting movement of this mass is expressed by the displacement q_{1z}.

The model illustrated in Fig. 3.26 is useful for sinusoidal or random excitation when the vibration intensity is included in the range from 0.5 m/s^2 up to 3 m/s^2. Numerical values of the model parameters are presented in Table 3.3. These values may be employed to describe the human behaviour of total body mass of 75 kg assuming that 73 % of the mass is resting on the seat. The parameter value m_2 shall be modified in order to obtain the total body mass for 55 kg or 90 kg.

The equation of motion for such a bio-mechanical model (Fig. 3.26) is determined on the basis of the modelling procedure discussed in Sect. 3.3. Hence, the system dynamics may be given as the set of ordinary differential equations:

$$\mathbf{M}_z\ddot{\mathbf{q}}_z + \mathbf{D}_z\dot{\mathbf{q}}_z + \mathbf{C}_z\mathbf{q}_z = \mathbf{F}_z \tag{3.62}$$

where $\mathbf{q}_z$ is the displacement vector of human body model, $\mathbf{M}_z$, $\mathbf{D}_z$, $\mathbf{C}_z$ are the inertia, damping and stiffness matrices, $\mathbf{F}_z$ is the vector of applied forces.

The vectors describing model displacements q_{1z}, q_{2z}, q_{3z}, q_{4z} and the force F_{1z} applied in vertical direction read:

$$\mathbf{q}_z = \begin{bmatrix} q_{1z} \\ q_{2z} \\ q_{3z} \\ q_{4z} \end{bmatrix}, \qquad \mathbf{F}_z = \begin{bmatrix} F_{1z} \\ 0 \\ 0 \\ 0 \end{bmatrix} \tag{3.63}$$

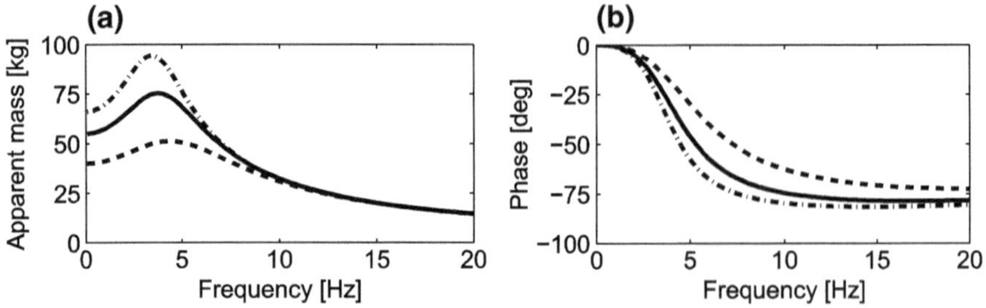

Fig. 3.27 Apparent mass and phase of the seated body under vertical vibration for total body masses: 55 kg (----), 75 kg (—) and 90 kg (-·-·-)

The diagonal inertia matrix $\mathbf{M}_z$ consists of elements that include individual body masses m_1, m_2, m_3 and m_4. The symmetric damping $\mathbf{D}_z$ and stiffness $\mathbf{C}_z$ matrices represented by the general Eqs. (3.8)–(3.11) are reorganised as follows:

$$\mathbf{D}_z = \begin{bmatrix} d_{12z} + d_{13z} & -d_{12z} & -d_{13z} & 0 \\ -d_{12z} & d_{12z} & 0 & 0 \\ -d_{13z} & 0 & d_{13z} + d_{34z} & -d_{34z} \\ 0 & 0 & -d_{34z} & d_{34z} \end{bmatrix} \quad (3.64)$$

$$\mathbf{C}_z = \begin{bmatrix} c_{12z} + c_{13z} & -c_{12z} & -c_{13z} & 0 \\ -c_{12z} & c_{12z} & 0 & 0 \\ -c_{13z} & 0 & c_{13z} + c_{34z} & -c_{34z} \\ 0 & 0 & -c_{34z} & c_{34z} \end{bmatrix} \quad (3.65)$$

For the purpose of evaluating the apparent mass of human body, the Eq. (3.62) is rearranged into the state-space form based on the modelling procedure shown in Sect 3.6. The resulting apparent mass and phase under vertical vibration are demonstrated in Fig. 3.27.

3.7.2 Apparent Mass Model Under Horizontal Vibration

In the papers [3, 25], various linear multi-degree of freedom models are discussed for the purpose of modelling the seated human body in horizontal directions. The model parameters are identified from averaged apparent mass values (modulus and phase) that are measured using 13 male subjects within the framework of the European research project VIBSEAT. The subjects with body masses between 62.2 and 103.6 kg and body heights between 174 and 196 cm are exposed to horizontal whole-body vibration of three different excitation intensities: $e_1 = 1\ \mathrm{m/s^2}$, $e_2 = 0.3\ \mathrm{m/s^2}$ and

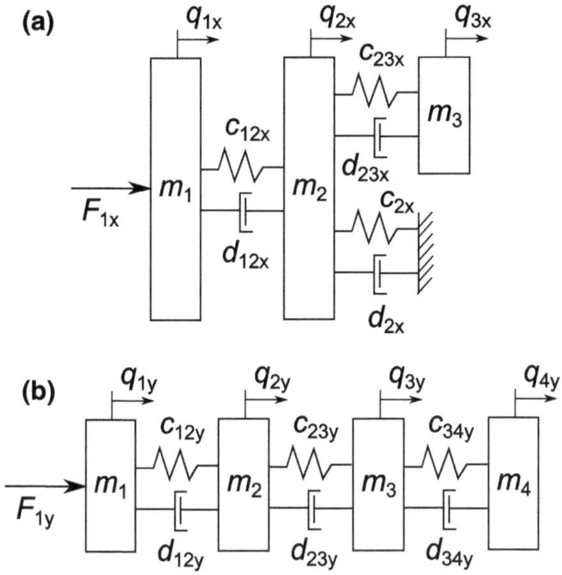

Fig. 3.28 Bio-mechanical model of the human body: longitudinal x-direction (**a**) and lateral y-direction (**b**) according to [3, 25]

$e_3 = 2\ \mathrm{m/s^2}$. The preferred model structures for longitudinal and lateral directions are shown in Fig. 3.28.

The apparent mass model parameters (Table 3.4) are identified by additional measurements under laboratory conditions. These data are applicable for modelling of horizontal vibration influence on the human body in sitting position when the cushioned seat is used and the backrest contact in the lumbar region is provided.

The system dynamics (Fig. 3.28) is determined in accordance to the modelling procedure presented in Sect. 3.3. The equations of motion are defined independently for the longitudinal x and lateral y vibration transmission as following:

$$\mathbf{M}_x\ddot{\mathbf{q}}_x + \mathbf{D}_x\dot{\mathbf{q}}_x + \mathbf{C}_x\mathbf{q}_x = \mathbf{F}_x, \qquad \mathbf{M}_y\ddot{\mathbf{q}}_y + \mathbf{D}_y\dot{\mathbf{q}}_y + \mathbf{C}_y\mathbf{q}_y = \mathbf{F}_y \tag{3.66}$$

where $\mathbf{q}_x$ and $\mathbf{q}_y$ the displacement vectors of human body models, $\mathbf{M}_x$, $\mathbf{M}_y$ $\mathbf{D}_x$, $\mathbf{D}_y$, $\mathbf{C}_x$, $\mathbf{C}_y$ are the corresponding inertia, damping and stiffness matrices, $\mathbf{F}_x$ and $\mathbf{F}_y$ are the vectors of applied forces.

The particular vectors defining the human body movements and applied forces are presented below:

$$\mathbf{q}_x = \begin{bmatrix} q_{1x} \\ q_{2x} \\ q_{3x} \end{bmatrix}, \quad \mathbf{F}_x = \begin{bmatrix} F_{1x} \\ 0 \\ 0 \end{bmatrix}, \quad \mathbf{q}_y = \begin{bmatrix} q_{1y} \\ q_{2y} \\ q_{3y} \\ q_{4y} \end{bmatrix}, \quad \mathbf{F}_y = \begin{bmatrix} F_{1y} \\ 0 \\ 0 \\ 0 \end{bmatrix} \tag{3.67}$$

Table 3.4 Values for the parameters of the model at different excitation intensities [3, 25]

Vertical direction	Excitation intensity, m/s^2	Mass, kg	Damping, Ns/m	Stiffness, N/m
Longitudinal x	e_1 0.3	m_1 11.2	d_{12x} 641.5	c_{12x} 73336
		m_2 55.6	d_{23x} 207.1	c_{23x} 3793
		m_3 10.9	d_{2x} 29.5	c_{2x} 0
Longitudinal x	e_2 1	m_1 15.1	d_{12x} 583.9	c_{12x} 50735
		m_3 53.4	d_{23x} 208.4	c_{23x} 1549
		m_3 13.2	d_{2x} 8.2	c_{2x} 0
Longitudinal x	e_3 2	m_1 16.9	d_{12x} 459.4	c_{12x} 39921
		m_2 51.1	d_{23x} 226.6	c_{23x} 788.1
		m_3 14.2	d_{2x} 5.5	c_{2x} 0
Lateral y	e_1 0.3	m_1 12.9	d_{12y} 865.4	c_{12y} 25090
		m_2 48.3	d_{23y} 58.4	c_{23y} 6925
		m_3 10.9	d_{34y} 63.7	c_{34y} 417.9
		m_4 13.1		
Lateral y	e_2 1	m_1 15.9	d_{12y} 662.8	c_{12y} 17977
		m_2 40.8	d_{23y} 70.2	c_{23y} 5117
		m_3 12.4	d_{34y} 60.2	c_{34y} 269.8
		m_4 12.8		
Lateral y	e_3 2	m_1 17.5	d_{12y} 543.9	c_{12y} 12168
		m_2 40.9	d_{23y} 61	c_{23y} 4205
		m_3 12	d_{34x} 156.7	c_{34y} 76.8
		m_4 14.6		

The diagonal inertia matrices $\mathbf{M}_x$ and $\mathbf{M}_y$ consist of the masses m_{1x}, m_{2x}, m_{3x} and m_{1y}, m_{2y}, m_{3y}, m_4 that occur in the human body models presented in Fig. 3.28. The damping matrices $\mathbf{D}_x$ and $\mathbf{D}_y$ described by Eqs. (3.8)–(3.11) take the following forms:

$$\mathbf{D}_x = \begin{bmatrix} d_{12x} & -d_{12x} & 0 \\ -d_{12x} & d_{2x} + d_{12x} + d_{23x} & -d_{23x} \\ 0 & -d_{23x} & d_{23z} \end{bmatrix} \quad (3.68)$$

$$\mathbf{D}_y = \begin{bmatrix} d_{12y} & -d_{12y} & 0 & 0 \\ -d_{12y} & d_{12y} + d_{23y} & -d_{23y} & 0 \\ 0 & -d_{23y} & d_{23y} + d_{34y} & -d_{34y} \\ 0 & 0 & -d_{34y} & d_{34y} \end{bmatrix} \quad (3.69)$$

The stiffness matrices $\mathbf{C}_x$ and $\mathbf{C}_y$ are determined analogously as follows:

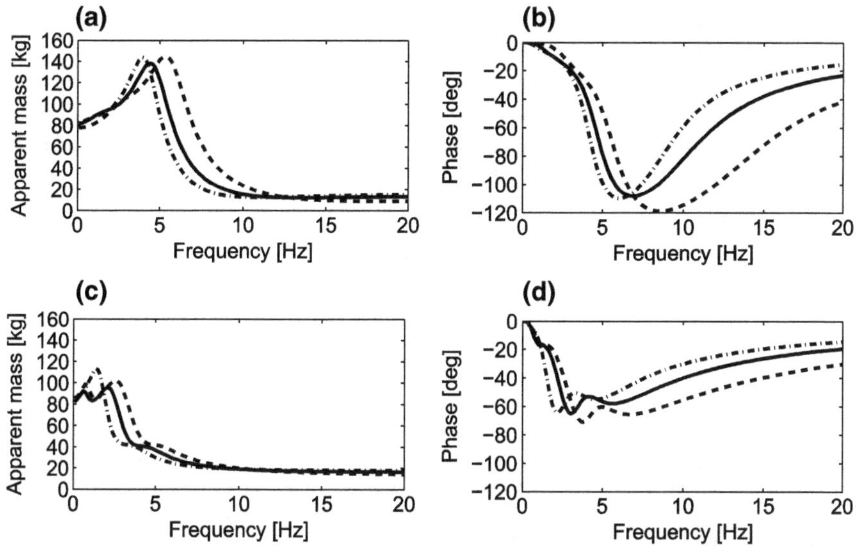

Fig. 3.29 Apparent mass and phase of the seated body under horizontal vibration: longitudinal direction (**a**) and (**b**), lateral direction (**c**) and (**d**) for excitation intensities: $e_1 = 0.3\ \mathrm{m/s^2}$ (----), $e_2 = 1\ \mathrm{m/s^2}$ (——), $e_3 = 2\ \mathrm{m/s^2}$ (-·-·-)

$$\mathbf{C}_\mathrm{x} = \begin{bmatrix} c_{12\mathrm{x}} & -c_{12\mathrm{x}} & 0 \\ -c_{12\mathrm{x}} & c_{2\mathrm{x}} + c_{12\mathrm{x}} + c_{23\mathrm{x}} & -c_{23\mathrm{x}} \\ 0 & -c_{23\mathrm{x}} & c_{23\mathrm{z}} \end{bmatrix} \tag{3.70}$$

$$\mathbf{C}_\mathrm{y} = \begin{bmatrix} c_{12\mathrm{y}} & -c_{12\mathrm{y}} & 0 & 0 \\ -c_{12\mathrm{y}} & c_{12\mathrm{y}} + c_{23\mathrm{y}} & -c_{23\mathrm{y}} & 0 \\ 0 & -c_{23\mathrm{y}} & c_{23\mathrm{y}} + c_{34\mathrm{y}} & -c_{34\mathrm{y}} \\ 0 & 0 & -c_{34\mathrm{y}} & c_{34\mathrm{y}} \end{bmatrix} \tag{3.71}$$

Thus, the Eqs. (3.66) are transformed again into the state-space models by means of the modelling procedure that have been discussed in Sect. 3.6. The simulated apparent mass and phase of the human body exposed to horizonal vibration in different directions are illustrated in Fig. 3.29.

3.8 Conclusions

The mathematical description presented in this chapter contributes the basement for modelling a mechanical analogue of the human exposed to whole-body vibration. The human body is represented by a discrete mechanical system with a large number of the degrees of freedom. The individual body masses are combined using typical visco-elastic elements. A novelty of the given model lies in a configurable mechanical structure that may be employed for the sake of selecting the vibro-isolation properties of vibration reduction systems. A suitable choice of the system dynamic characteristics might be performed for a specific design of the vibration isolator that oscillates in accordance with a specific signal representing the work of unique machinery. The possibility of modifying the system characteristics is achieved by changing the model parameters of a typical suspension components. Moreover, an exemplary model of the horizontal seat suspension is presented and the system dynamics is investigated experimentally for the selected input vibration.

References

1. Directive 2002/44/EC of the European Parliament and of the Council. (2002). *On the minimum health and safety requirements regarding the exposure of workers to the risks arising from physical agents (vibration)*, Official Journal of the European Communities (pp. 13–18).
2. Rutzel, S., Hinz, B., & Wolfel, H. B. (2006). Modal description-A better way of characterizing human vibration behavior. *Journal of Sound and Vibration, 298*, 810–823.
3. Stein, G. J., Muka, P., Chmurny, R., Hinz, B., & Bluthner, R. (2007). Measurement and modelling of x-direction apparent mass of the seated human body-Cushioned seat system. *Journal of Biomechanics, 40*, 1493–1503.
4. Toward, M., & Griffin, J. (2011). The transmission of vertical vibration through seats: Influence of the characteristics of the human body. *Journal of Sound and Vibration, 330*, 6526–6543.
5. Boer, L. J. A. (2009). *Nonlinear dynamic behavior of a conical spring with top mass, Master's thesis*. Eindhoven: Eindhoven University of Technology.
6. Rodriguez, E., Paredes, M., & Sartor, M. (2006). Analytical behavior law for a constant pitch conical compression spring. *Journal of Mechanical Design, Transactions of the ASME, 128*, 1352–1356.
7. Mansfield, N. J. (2005). *Human response to vibration*. Boca Raton: CRC Press.
8. Gunston, T. P., Rebelle, J., & Griffin, M. J. (2004). A comparison of two methods of simulating seat suspension dynamic performance. *Journal of Sound and Vibration, 278*, 117–134.
9. Titurus, B., Bois, J., Lieven, N., & Hansford, R. (2010). A method for the identification of hydraulic damper characteristics from steady velocity inputs. *Mechanical Systems and Signal Processing, 24*, 2868–2887.
10. Czop, P., & Slawik, D. (2011). A high-frequency first-principle model of ashock absorber and servo-hydraulic tester. *Mechanical Systems and Signal Processing, 25*, 1937–1955.
11. Ferreira, C., Ventura, P., Morais, R., Valente, A., Neves, C., & Reis, M. (2009). Sensing methodologies to determine automotive damper condition under vehicle normal operation. *Sensors and Actuators A: Physical, 156*, 237–244.
12. Idelchik, I. E. (2008). *Handbook of hydraulic resistance-4th edition revised and augmented research*. Moscow: Institute for Gas Purification.
13. Tarnowski, W. (2001). *Simulation and optimisation in matlab (in Polish)*. Sopot: Intergraf S.C.
14. Borello, L., & Dalla Vedova M. D. L. (2014). Dry friction discontinuous computational algorithms. *International Journal of Engineering and Innovative Technology, 3*(8), 1–8.
15. Maciejewski, I., Kiczkowiak, T., & Krzyzynski, T. (2011). Application of the Pareto-optimal approach for selecting dynamic characteristics of seat suspension systems. *Vehicle System Dynamics, 49*(12), 1929–1950.
16. Zipser, L., Richter, L., & Lange, U. (2001). Magnetorheological fluids for actuators. *Sensors and Actuators A: Physical, 92*, 318–325.

17. Kwok, M. M., Ha, Q. P., Nguyen, T. H., Li, J., & Samali, B. (2006). A novel hysteretic model for magnetorheological fluid dampers and parameter identification using particle swarm optimization. *Sensors and Actuators A*, *132*, 441–451.
18. Maciejewski, I., Krzyzynski, T., & Meyer, L. (2014). Control system synthesis of seat suspensions used for protection of working machine operators. *Vehicle System Dynamics*, *52*(11), 1355–1371.
19. Jezierski, E. (2006). *Dynamika robotów*. Warszawa: Wydawnictwa Naukowo-Techniczne.
20. Sarosi, J., Biro, I., Nemeth, J., & Cveticanin, L. (2015). Dynamic modeling of a pneumatic muscle actuator with two-direction motion. *Mechanism and Machine Theory*, *85*, 25–34.
21. Bogacki, P., & Shampine, L. F. (1989). A 3(2) pair of Runge-Kutta formulas. *Applied Mathematics Letters*, *2*(4), 321–325.
22. Griffin, M. J. (1996). *Handbook of human vibration*. London: Elsevier Academic Press.
23. International Organization for Standardization. (2001). *Mechanical vibration and shock-Range of idealized values to characterize seated-body biodynamic response under vertical vibration*. ISO 5982, Genewa.
24. Maeda, S., & Mansfield, N. J. (2005). Comparison of the apparent mass during exposure to whole-body vertical vibration between Japanese subjects and ISO 5982 standard. *Industrial Health*, *43*, 436–440.
25. Stein, G. J., Muka, P., Hinz, B., & Bluthner, R. (2009). Measurement and modelling of the y-direction apparent mass of sitting human body-cushioned seat system. *Journal of Sound and Vibration*, *322*, 454–474.

Chapter 4
Vibro-isolation Criteria

4.1 Nomenclature

$CSD_{\ddot{q}_{si}\ddot{q}_{1i}}(2\pi f)$	cross-spectral density of two different signals: acceleration of the input vibration $\ddot{q}_{si}$ and acceleration of the isolated body $\ddot{q}_{1i}$
$H_{hi}(s)$, $H_{li}(s)$	transfer functions of the filters limiting the scope of the analysed vibration frequency
$H_{ti}(s)$, $H_{si}(s)$	transfer functions of the filters shaping the characteristics of frequency weightings
$H_{kfi}(s)$	transfer function of the total filter according to the frequency weightings W_k i W_f
$H_{dmi}(s)$	transfer function of the total filter according to the frequency weightings W_d i W_m
$PSD_{\ddot{q}_{si}\ddot{q}_{si}}(2\pi f)$	power spectral density of the excitation acceleration $\ddot{q}_{si}$
$Q_1, \ldots, Q_6$	resonant quality factors specified for different directions of the vibration exposure ($\mathrm{i} = x, y, z$)
$T_i(2\pi f)$	transmissibility function for selected direction of the vibration exposure ($\mathrm{i} = x, y, z$)
TFE_i	frequency-weighted transmissibility factor for selected direction of the vibration exposure ($\mathrm{i} = x, y, z$)
$TR_i(2\pi f_r)$	maximum amplification of the transmissibility function at resonance of vibration reduction system and for selected direction of the vibration exposure ($\mathrm{i} = x, y, z$)
VDV	total vibration dose value, $\mathrm{m/s^{1.75}}$
VDV_{1iw}	vibration dose value for selected direction of the vibration exposure ($\mathrm{i} = x, y, z$), $\mathrm{m/s^{1.75}}$
VDV_{8h}	vibration dose value during a working day, $\mathrm{m/s^{1.75}}$
$(a_{1iw})_{RMS}$	root mean square value of the frequency-weighted acceleration for a single axis of the vibration transmission ($\mathrm{i} = x, y, z$), $\mathrm{m/s^2}$

T. Krzyzynski et al., *Modelling and Control Design of Vibration Reduction Systems*, https://doi.org/10.1007/978-3-030-03047-6_4

a_{8h}	root mean square acceleration during a working day, m/s^2
a_{RMS}	total value of weighted root mean square acceleration, m/s^2
$(a_{1i})_{RMS}(2\pi f_r)$	root mean square value of the suspended body acceleration at resonance of vibration reduction system and for selected axis of the vibration transmission ($i = x, y, z$), m/s^2
$(a_{si})_{RMS}(2\pi f_r)$	root mean square value of the input acceleration at resonance of vibration reduction system and for selected axis of the vibration transmission ($i = x, y, z$), m/s^2
f	frequency, Hz
$f_1, ..., f_6$	cut-off frequencies of the weighting filters, Hz
f_r	resonance frequency, Hz
q_{1i}	displacement of the isolated body for selected direction of the vibration exposure ($i = x, y, z$), m
$\ddot{q}_{1iw}$	frequency-weighted acceleration of the human body model for selected direction of the vibration exposure ($i = x, y, z$), m/s^2
q_{si}	displacement of the input vibration ($i = x, y, z$), m
s	Laplace variable
s_{ti}	suspension travel for selected direction of the vibration exposure ($i = x, y, z$), m
t_{exp}	daily duration of exposure to the vibration, s
t_k	computation time, s.

4.2 Introduction

Effects of the vibration influence is usually considered depending on a location of the vibration transmission and a position of the human body during operation. Whole-body vibration occurs when the body is supported by a structure which vibrates significantly. There are two principal work opportunities, i.e. sitting on a vibrating seat or standing on a vibrating floor. Local vibration occurs when the limbs are in contact with operating equipments that generate oscillatory forces. Both whole-body and local vibration can be injurious to health [1].

The vibration experienced in the typical working machines usually occurs at frequencies from 1 to 20 Hz [2–4]. Such a phenomenon is very undesirable, because the human body has a number of natural frequencies in this range. For example, the natural frequencies of the head are equal to 4 and 25 Hz, the jaw has the natural frequency within the range 6–8 Hz, natural frequencies of the internal organs of the chest are between 5 and 8 Hz, and the internal organs of the abdominal cavity possess the frequency range 4.5–10 Hz [5]. Even low excitation amplitudes at these frequencies can cause resonant vibration of the individual human body parts and organs. The consequences for operators are: loss of concentration, tiredness and decrease of the effectiveness of the work being performed.

In the following chapter, the human exposure to whole-body vibration is discussed that takes into account the resonances of human body parts and organs using the frequency-weighting method. Based on this method, the vibro-isolation criteria are formulated in order to evaluate the vibration reduction system effectiveness for different working conditions. Then dynamic properties of the exemplary horizontal seat suspension are determined using a laboratory method for measuring and evaluating the system effectiveness in reducing the whole-body vibration.

4.3 Evaluation of the Human Exposure to Whole-Body Vibration

The principal method of evaluating the human exposure to vibration includes the measurements of the frequency-weighted acceleration [6]. Because the risk of vibration is not equal at all frequencies, the frequency weightings are utilised in order to distinguish the human sensitivity at different vibration frequencies. These curves define the gain factors to be multiplied by the vibration magnitude at each frequency. For instance, the most dangerous frequency range of external vertical vibration for the seated human body is between 4 and 8 Hz, because there are several body eigenfrequencies [7, 8]. As a result, the frequency weighting should demonstrate the highest factor in the resonance frequency range which decreases beyond that range.

The frequency weighting of acceleration signal shall be done in accordance with ISO-2631 [6]. Referring to this standard, principal frequency weightings recommended for the various directions and their applications are presented in Table 4.1. In order to perform a frequency weighting of the measured vibration acceleration, the component filter sections should be defined as the analogue transfer functions. Band-limiting is performed using 'high-pass' and 'low-pass' filters, and other component parts are archived using 'acceleration–velocity transition' and 'upward step' filters. This can be expressed in the complex domain as follows [9]:

- high-pass filter (Butterworth characteristics)

$$H_{\mathrm{hi}}(s) = \frac{s^2}{s^2 + \frac{2\pi f_1}{Q_1} s + (2\pi f_1)^2}, \qquad \mathrm{i} = x, y, z \tag{4.1}$$

- low-pass filter (Butterworth characteristics)

$$H_{\mathrm{li}}(s) = \frac{(2\pi f_2)^2}{s^2 + \frac{2\pi f_2}{Q_2} s + (2\pi f_2)^2}, \qquad \mathrm{i} = x, y, z \tag{4.2}$$

Table 4.1 Principal frequency weightings for the exposure evaluation according to ISO-2631 [6]

Human body position	Measuring place	Measuring direction	Frequency weighting
Health risk assessment			
Seating position	Seat surface	z x, y	W_k W_d
Comfort assessment			
Seating position	Seat surface	z x, y	W_k W_d
Seating position	Vibrating platform	x, y, z	W_k
Standing position	Vibrating platform	z x, y	W_k W_d
Non-specified position	Floor in buildings	x, y, z	W_m
Probability of motion sickness			
Seating or standing position	Seat surface or vibrating basement	z	W_f

- acceleration–velocity transition filter (a proportionality to the acceleration at lower frequencies, a proportionality to the velocity at higher frequencies)

$$H_{ti}(s) = \frac{\frac{(2\pi f_4)^2}{2\pi f_3} s + (2\pi f_4)^2}{s^2 + \frac{2\pi f_4}{Q_4} s + (2\pi f_4)^2}, \qquad i = x, y, z \tag{4.3}$$

- upward step filter (a proportionality to the jerk)

$$H_{si}(s) = \frac{s^2 + \frac{2\pi f_5}{Q_5} s + (2\pi f_5)^2}{s^2 + \frac{2\pi f_6}{Q_6} s + (2\pi f_6)^2}, \qquad i = x, y, z \tag{4.4}$$

where $s = j\omega/\omega_c$ is the Laplace variable, $j = \sqrt{-1}$ is the imaginary unit, $f_1, ..., f_6$ are the filter frequencies and $Q_1, ..., Q_6$ are the resonant quality factors specified for different directions of the vibration exposure ($i = x, y, z$). Parameters of the transfer functions described by Eqs. (4.1)–(4.4) are presented in Table 4.2.

The total weighting filters are defined as a combination of Eqs. (4.1–4.4). The transfer functions of such filters are defined as follows:

- total filter according to the frequency weightings W_k and W_f

$$H_{kfi}(s) = H_{hi}(s) \cdot H_{li}(s) \cdot H_{ti}(s) \cdot H_{si}(s), \qquad i = x, y, z \tag{4.5}$$

- total filter according to the frequency weightings W_d and W_m

$$H_{dmi}(s) = H_{hi}(s) \cdot H_{li}(s) \cdot H_{ti}(s), \qquad i = x, y, z \tag{4.6}$$

Table 4.2 Parameters of the transfer functions used for frequency weightings

	High-pass filter		Low-pass filter		Acceler.-velocity transition filter			Upward step filter			
	f_1, Hz	Q_1	f_2, Hz	Q_2	f_3, Hz	f_4, Hz	Q_4	f_5, Hz	Q_5	f_6, Hz	Q_6
W_k	0.4	0.707	100	0.707	12.5	12.5	0.63	2.37	0.91	3.3	0.9
W_d	0.4	0.707	100	0.707	2	2	0.63	–	–	–	–
W_m	0.79	0.707	100	0.707	5.68	5.68	0.5	–	–	–	–
W_f	0.08	0.707	0.63	0.707	$f_s/2$	0.25	0.86	0.06	0.8	0.1	0.8

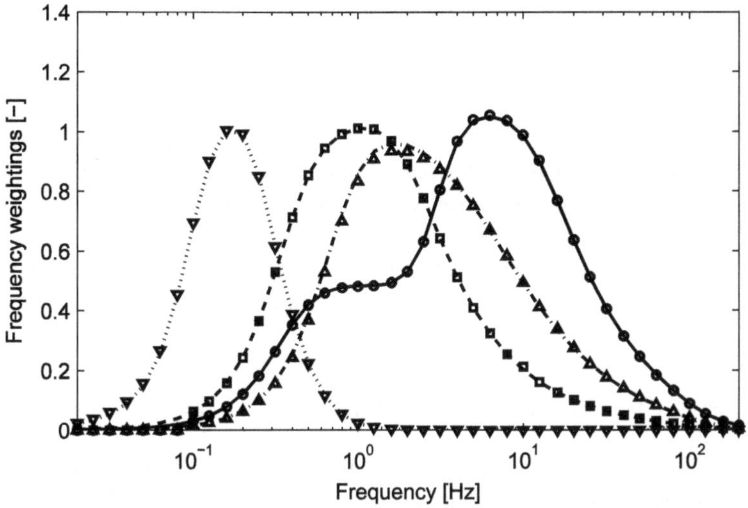

Fig. 4.1 Magnitudes of the total frequency-weighting filter according to frequency weightings: W_k (——), W_d (----), W_m (-·-·-), W_f (·······) and the principal values of the frequency weightings defined in ISO-2631: W_k (○), W_d (□), W_m (△), W_f (▽)

In Fig. 4.1, the absolute magnitude of the total frequency-weighting filter and the principal values of the frequency-weighting coefficient defined in ISO-2631 [6] are presented.

The running root mean square (RMS) method assumes occasional shocks and transient vibration [6]. The vibration magnitude is calculated on the basis of the frequency-weighted acceleration signal by means of the weighting filters (Eqs. 4.5–4.6) as follows:

$$(a_{1iw})_{RMS} = \sqrt{\frac{1}{t_k}\int_0^{t_k} \ddot{q}^2_{1iw}(t)dt}, \qquad i = x, y, z \tag{4.7}$$

where $\ddot{q}_{1iw}^2(t)$ is the second power of the isolated body acceleration measured along a selected axis of the Cartesian coordinate system (i = x, y, z) and t_k is the time of observation.

The fourth power vibration dose value (VDV) method is more sensitive to peaks than the running root mean square method by using the fourth power instead of the second power of the acceleration time history [6]. The vibration magnitude is defined using the following relation:

$$\mathrm{VDV}_{1iw} = \sqrt[4]{\int_0^{t_k} \ddot{q}_{1iw}^4(t)dt}, \qquad \mathrm{i} = x, y, z \tag{4.8}$$

where $\ddot{q}_{1iw}^4(t)$ is the fourth power of the isolated body acceleration measured for a specific direction (i = x, y, z). The units for VDV_{1iw} are metre per second to the power 1.75 ($\mathrm{m/s}^{1.75}$) and unlike the root mean square vibration magnitude $(a_{1iw})_{\mathrm{RMS}}$, it is a cumulative value that increases with the measurement time t_k [10]. Therefore, using this value it is important to know the period over which the value is measured.

According to the standard [6], the vibration total value of weighted root mean square acceleration, determined from vibration in three orthogonal coordinates, is calculated as follows:

$$a_{\mathrm{RMS}} = \sqrt{(k_x \cdot (a_{1xw})_{\mathrm{RMS}})^2 + \left(k_y \cdot (a_{1yw})_{\mathrm{RMS}}\right)^2 + (k_z \cdot (a_{1zw})_{\mathrm{RMS}})^2} \tag{4.9}$$

where $(a_{1xw})_{\mathrm{RMS}}$, $(a_{1yw})_{\mathrm{RMS}}$, $(a_{1zw})_{\mathrm{RMS}}$ are the weighted root mean square accelerations with respect to the orthogonal axes x, y, z.

In a similar way, the total vibration dose value can be calculated as a sum of the three orthogonal axis values VDV_{1xw}, VDV_{1xw}, VDV_{1xw} in the following form:

$$\mathrm{VDV} = \sqrt[4]{(k_x \cdot \mathrm{VDV}_{1xw})^4 + \left(k_y \cdot \mathrm{VDV}_{1yw}\right)^4 + (k_z \cdot \mathrm{VDV}_{1zw})^4} \tag{4.10}$$

where k_x, k_y, k_z are the multiplying factors used for the exposure assessment. Their values in the case of horizontal directions are equal to $k_x = k_y = 1.4$, while for a vertical direction it is equal to $k_z = 1$.

The measurement results obtained by using the frequency-weighting method can be then analysed with respect to the total daily duration of exposure to the vibration from working machines used. Daily vibration exposure depends on both the level of vibration and the duration of exposure as follows:

$$a_{8h} = a_{\mathrm{RMS}}\sqrt{\frac{t_{\mathrm{exp}}}{t_h}} \tag{4.11}$$

$$\mathrm{VDV}_{8h} = \mathrm{VDV}\sqrt{\frac{t_{\mathrm{exp}}}{t_h}} \tag{4.12}$$

where a_{8h} is the root mean square acceleration during a working day, VDV_{8h} is the vibration dose value during a working day, t_{exp} is the daily duration of exposure to the vibration and t_h is the reference duration of eight hours.

Directive 2002/44/EC defines the minimum health and safety requirements regarding the exposure of workers to the risks arising from vibrations. This 'Vibration Directive' [11] specifies an exposure action value and an exposure limit value above which working machine operators must not be exposed:

- the root mean square acceleration a_{8h} standardised to an eight-hour reference period
  - the daily exposure action value $(a_{8h})_a = 0.5\ \mathrm{m/s^2}$,
  - the daily exposure limit value $(a_{8h})_l = 1.15\ \mathrm{m/s^2}$,
- the vibration dose value VDV_{8h} standardised to an eight-hour reference period
  - the daily exposure action value $(VDV_{8h})_a = 9.1\ \mathrm{m/s^{1.75}}$,
  - the daily exposure limit value $(VDV_{8h})_l = 21\ \mathrm{m/s^{1.75}}$.

Figures 4.2 and 4.3 illustrate a simple alternative method for looking up daily exposures or partial vibration exposures. There are three levels of the human exposure that can be distinguished:

- if $a_{RMS} < (a_{8h})_a$ and $VDV < (VDV_{8h})_a$, then a low exposure to vibration is observed,

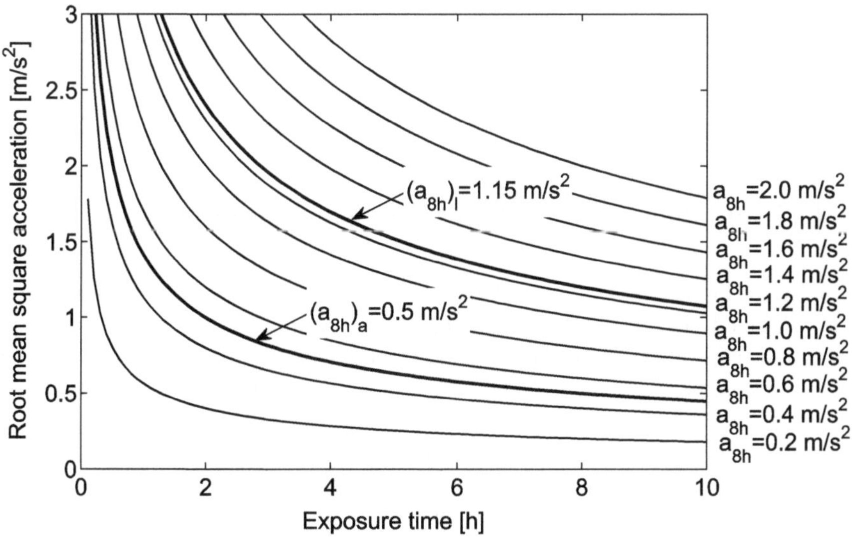

Fig. 4.2 Boundary values of the root mean square acceleration for different exposures to vibration during an eight-hour working day

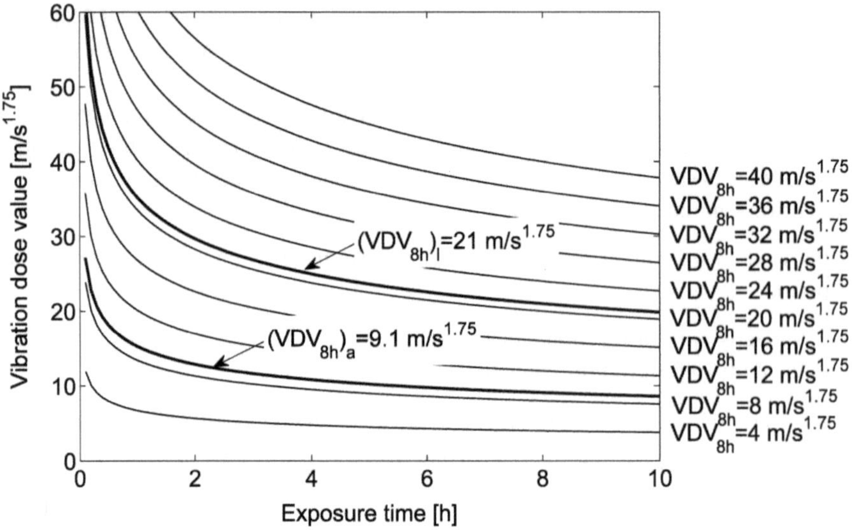

Fig. 4.3 Boundary values of the vibration dose value for different exposures to vibration during an eight-hour working day

- if $(a_{8h})_a \leq a_{RMS} < (a_{8h})_l$ and $(VDV_{8h})_a \leq VDV < (VDV_{8h})_l$, then a significant exposure to vibration is obtained,
- if $a_{RMS} \geq (a_{8h})_l$ and $VDV \geq (VDV_{8h})_l$, then an unacceptable exposure is achieved.

Even if the root mean square acceleration a_{8h} or the vibration dose value VDV_{8h} is located in the range of a low exposure to vibration, then safe working conditions cannot be obtained [11].

4.4 Evaluation of the System Vibro-isolation Properties

A selection process of the system dynamic properties is difficult due to a presence of the conflicted vibro-isolation criteria [12]. In the case of an ideal system isolation, vibration amplitude q_{1i} of the isolated body is equal to zero (Fig. 4.4a) and the human vibrations are perfectly dissipated by a suspension system. Unfortunately, in such a situation the relative displacements $q_{1i} - q_{si}$ of suspension system are considerably large and they are the same as the input vibration q_{si}. In an extreme case, the recommended deflection limitation requires the system relative displacement equal to zero (Fig. 4.4b). Then the isolated body oscillates with the vibration amplitude q_{1i} equal to the input vibration q_{si} and endanger operator's health.

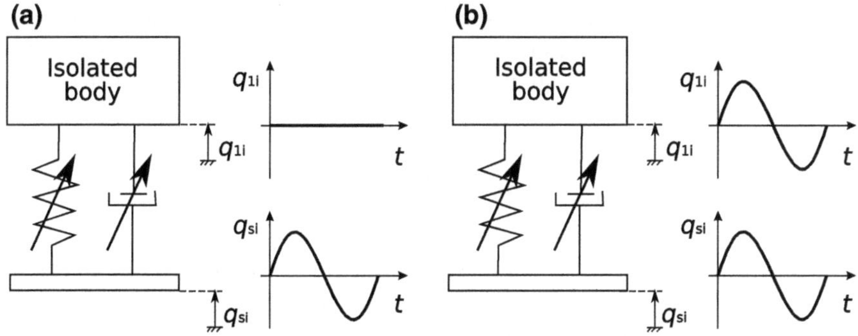

Fig. 4.4 Conflicted requirements for vibro-isolating properties: ideal isolation of the body against vibration (**a**) and ideal limitation of the relative displacement of suspension system (**b**)

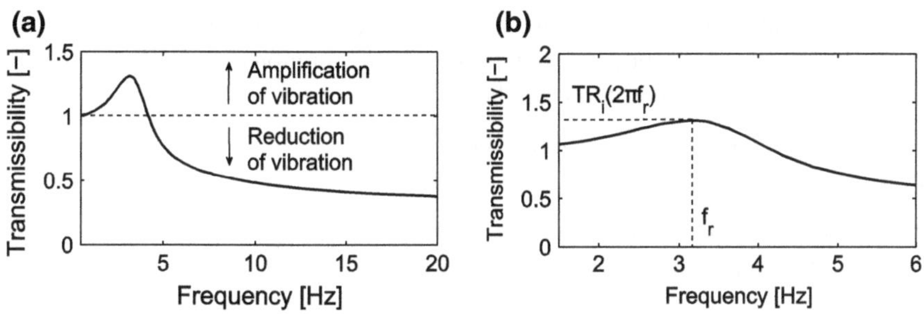

Fig. 4.5 Transmissibility function of exemplary vibration reduction system (**a**) and maximum amplification of the transmissibility function at resonance (**b**)

In order to evaluate the quality performance of vibration reduction system, the transmissibility function is commonly used. This function can be evaluated using the following expression [1]:

$$T_{\mathrm{i}}(2\pi f) = \frac{\mathrm{CSD}_{\ddot{q}_{si}\ddot{q}_{1i}}(2\pi f)}{\mathrm{PSD}_{\ddot{q}_{si}\ddot{q}_{si}}(2\pi f)}, \qquad \mathrm{i} = x, y, z \tag{4.13}$$

where $\mathrm{CSD}_{\ddot{q}_{si}\ddot{q}_{1i}}(2\pi f)$ is the cross-spectral density of two different signals: acceleration of the input vibration $\ddot{q}_{si}(t)$ and acceleration of the isolated body $\ddot{q}_{1i}(t)$ that are measured for a specific direction of the vibration transmission ($\mathrm{i} = x, y, z$), $\mathrm{PSD}_{\ddot{q}_{si}\ddot{q}_{si}}(2\pi f)$ is the power spectral density of the excitation acceleration $\ddot{q}_{si}$, f is the frequency in Hz. The transmissibility function of exemplary vibration reduction system is shown in Fig. 4.5a.

To determine the maximum amplification of vibration, the transmissibility at resonance should be evaluated by means of the damping test. According to the ISO standard [4], the suspension shall be excited by a sinusoidal vibration of linearly increasing frequency from 0.5 to 2 times than the expected resonance frequency of the system. Amplitude of the peak-to-peak displacement should be constant and equal to 40% of the total suspension stroke. For such defined test conditions, the maximum amplification of the transmissibility function at resonance can be calculated using the following relationship:

$$\mathrm{TR_i}(2\pi f_\mathrm{r}) = \frac{(a_{1\mathrm{i}})_\mathrm{RMS}(2\pi f_\mathrm{r})}{(a_\mathrm{si})_\mathrm{RMS}(2\pi f_\mathrm{r})}, \qquad \mathrm{i} = x, y, z \tag{4.14}$$

where $(a_{1\mathrm{i}})_\mathrm{RMS}(2\pi f_\mathrm{r})$ is the root mean square value of the isolated body acceleration and $(a_\mathrm{si})_\mathrm{RMS}(2\pi f_\mathrm{r})$ is the root mean square value of the input acceleration that are measured at the resonance frequency f_r of vibration reduction system for a specific direction ($\mathrm{i} = x, y, z$). Graphical representation of the maximum amplification of the transmissibility function at resonance $\mathrm{TR_i}(2\pi f_\mathrm{r})$ is presented in Fig. 4.5b.

The vibro-isolation criteria, that correspond to the conflicted system requirements, provide a simple numerical assessment of the suspension system efficiency. For the purpose of shaping the vibro-isolation properties of vibration reduction systems, the following vibro-isolation criteria are chosen:

- frequency-weighted transmissibility factor $\mathrm{TFE_i}$,
- suspension travel s_ti.

At first, the frequency-weighted transmissibility factor is defined in order to evaluate the effectiveness of vibration reduction system:

$$\mathrm{TFE_i} = \frac{(a_{1\mathrm{iw}})_\mathrm{RMS}}{(a_\mathrm{siw})_\mathrm{RMS}}, \qquad \mathrm{i} = x, y, z \tag{4.15}$$

where $(a_{1\mathrm{iw}})_\mathrm{RMS}$ is the frequency-weighted root mean square value of the suspended body acceleration for specific direction ($\mathrm{i} = x, y, z$) and $(a_\mathrm{siw})_\mathrm{RMS}$ is the frequency-weighted root mean square value of the input acceleration. In order to designate such a vibro-isolation criterion (Eq. 4.15), the frequency weightings specified in ISO-2631 [6] are employed. Principal frequency weightings recommended for various directions of the vibration transmission and different working postures are presented in Table 4.1.

If the value of $\mathrm{TFE_i}$ factor equals 1, it means that the vibration discomfort of drivers is the same as seating or standing on the base without any vibration isolation system. If the $\mathrm{TFE_i}$ factor is higher than 1, vibration amplitudes are amplified by the suspension system. However for the $\mathrm{TFE_i}$ factor lower than 1, the desired effect of vibration isolation is achieved by means of the suspension system [13].

For further approach, the suspension travel is considered as the second vibro-isolation criterion [12]:

$$s_\mathrm{ti} = \max_{t\in[0,t_\mathrm{k}]} \left(q_{1\mathrm{i}}(t) - q_\mathrm{si}(t)\right) - \min_{t\in[0,t_\mathrm{k}]} \left(q_{1\mathrm{i}}(t) - q_\mathrm{si}(t)\right), \quad \mathrm{i} = x, y, z \tag{4.16}$$

where $q_{1i}(t)$ is the displacement of suspended body for specific direction (i $= x, y, z$), $q_{si}(t)$ is the displacement of input vibration and t is the present moment of time and t_k is the time of observation.

To determine such a specific vibro-isolation criterion (Eq. 4.16), it is necessary to store the relative displacement $q_{1i}(t) - q_{si}(t)$ of suspension system during operation. Based on such a signal, the maximum deflection (rebound) of the system can be calculated and the obtained values should be kept as small as possible. It provides the desired contact of drivers with steering devices.

4.5 Vibro-isolation Properties of the Horizontal Seat Suspension

The experimental investigations are performed using the same test set-up that has been described in Sect. 3.5.3. The electrohydraulic shaker is utilised as an actuator that provides a random vibration source for testing and analysis. Two high output linear accelerometers are installed in order to measure the vibration of the shaker and the suspended body. Additionally, the cable extension position sensor is installed for measuring the relative displacement of the suspension system. PC-based data acquisition system is used as a combination of modular hardware, application software and a computer to take measurements. The measurement results are reported for a mass-loaded suspension system mechanism. In order to compare the simulation and measurement results, a seat cushion is removed and a metal mass load is mounted rigidly on the suspension system as shown in Fig. 4.6.

Fig. 4.6 Experimental set-up for the evaluation of the system vibro-isolation properties

In this section, the object of vibration isolation is assumed as the lumped mass body. This does not reflect all the properties of a real, complex vibration isolated system such as a human operator body seated on the cushion. It is obvious that the dynamics of the cushion and the operator can affect the seat suspension performance. However, such assumption makes the experimental investigations repeatable and, in consequence, the measured and computed results may be compared more precisely. For example, differences in the posture of seated person can cause a 10% difference in test results [4]. Moreover, the height and mass distribution of a tested person cause the disturbances in test results, making such investigations hardly ever repeatable.

For the purpose of evaluating the system vibro-isolation properties, the excitation signal similar to white, band-limited noise in the range of frequency 0.5–10 Hz are applied at three different excitation intensities: WN1x, WN3x and WN5x. The power spectral densities of these excitation signals have been shown in Fig. 3.20 (see Sect. 3.5.3). The obtained transmissibility functions of the horizontal suspension system for the different values of mass, i.e. 30, 55 and 80 kg, are presented in Figs. 4.7, 4.8 and 4.9.

Based on a qualitative assessment of the obtained research results, the operation of a seat suspension system is found to be ineffective in the low-frequency range, i.e. at frequencies of up to about 4 Hz. The conventional passive system increases the vibration amplitudes in a frequency range around the natural frequency of the system, although its effectiveness increases with over-resonant frequencies. Exception of this rule is the system dynamic behaviour obtained for the low excitation intensity

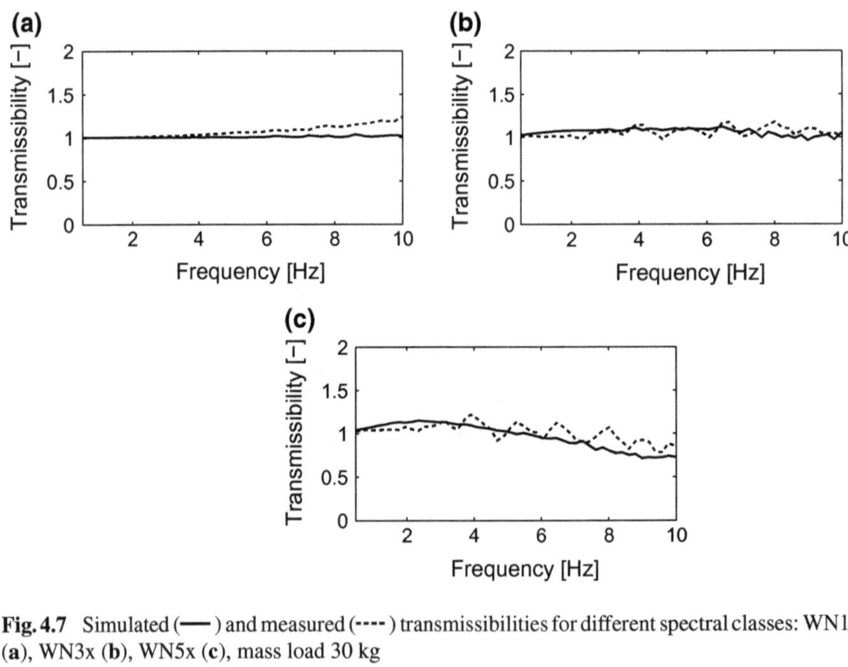

Fig. 4.7 Simulated (—) and measured (----) transmissibilities for different spectral classes: WN1x (**a**), WN3x (**b**), WN5x (**c**), mass load 30 kg

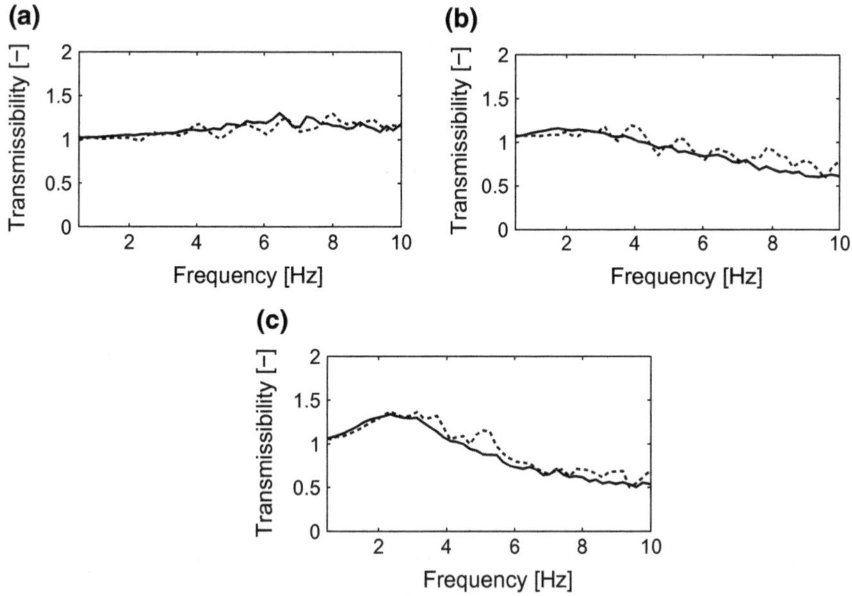

Fig. 4.8 Simulated (—) and measured (----) transmissibilities for different spectral classes: WN1x (**a**), WN3x (**b**), WN5x (**c**), mass load 55 kg

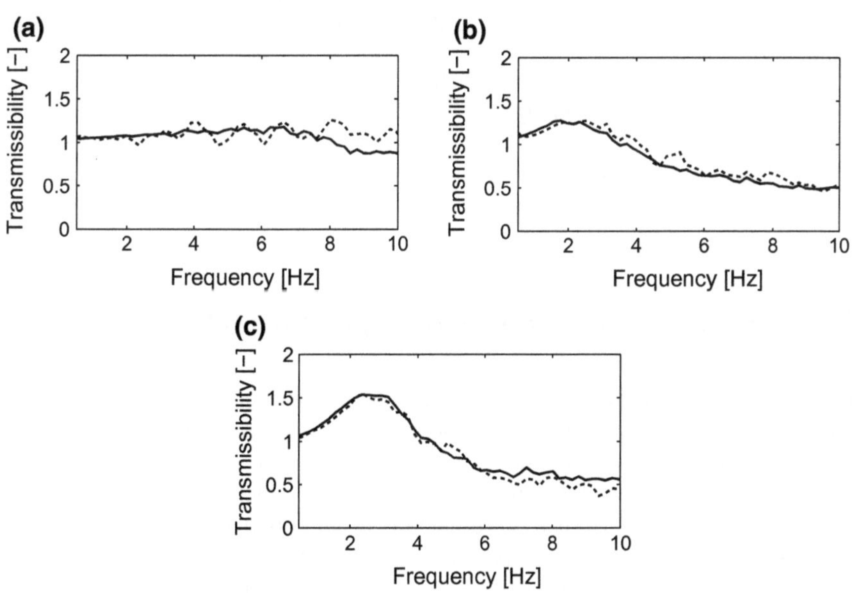

Fig. 4.9 Simulated (—) and measured (----) transmissibilities for different spectral classes: WN1x (**a**), WN3x (**b**), WN5x (**c**), mass load 80 kg

Table 4.3 Simulated and measured transmissibility factors, suspension travels and relative errors of the horizontal seat suspension for excitation signals: WN1x, WN3x and WN5x

		Simulation		Measurement		Relative error	
Suspended mass (kg)	Input vibration	TFE_x factor	s_{tx}, mm	TFE_x Factor	s_{tx}, mm	δ_{TFE_x}, %	$\delta_{s_{tx}}$, %
30	WN1x	1.004	2.3	1.033	2.1	6.0	9.5
	WN3x	1.070	16.1	1.039	15.2	2.9	5.9
	WN5x	1.086	23.6	1.056	22.6	2.8	4.4
55	WN1x	1.071	9.9	1.058	9.1	1.2	8.8
	WN3x	1.074	24.2	1.061	25.9	1.2	6.5
	WN5x	1.168	31.4	1.186	32.3	1.5	2.7
80	WN1x	1.077	15.7	1.061	17.1	1.5	8.1
	WN3x	1.108	30.4	1.121	31.6	1.1	3.8
	WN5x	1.272	37.6	1.252	36.7	1.6	2.5

$1.02\,m/s^2$ (spectral class WN1x). In this case, the vibration reduction is rather poor, because the friction prevents a proper functioning of the suspension system.

In addition of to qualitative assessment of the transmissibility functions, numerical values of the chosen criteria, i.e. the frequency-weighted transmissibility factor TFE_x and suspension travel s_{tx}, are compared in Table 4.3. Relative errors between measurement and simulation results are determined according to the following relations:

$$\delta_{TFE_x} = \frac{|(TFE_x)_s - (TFE_x)_m|}{(TFE_x)_m}, \qquad \delta_{s_{tx}} = \frac{|(s_{tx})_s - (s_{tx})_m|}{(s_{tx})_m} \tag{4.17}$$

where $(TFE_z)_s$ and $(s_{tz})_s$ are the simulated vibro-isolation criteria, $(TFE_z)_m$ and $(s_{tz})_m$ are the measurement results.

4.6 Laboratory Evaluation of the Operator Seat Vibration

The International Standard ISO-7096 [4] specifies a laboratory method for measuring and evaluating the effectiveness of the seat suspension in reducing the whole-body vibration transmitted to the operator of working machines. The experimental set-up for laboratory evaluation of the operator seat vibration, which meets all requirements for the person posture on a seat, is presented in Fig. 4.10b. In Fig. 4.10a, the physical model of the horizontal seat suspension with seated human body is shown. The mathematical models of the basic forces in the system, i.e. the spring force F_{cx1}, the forces from end-stop buffers F_{cx2}, the damping force of hydraulic shock-absorber F_{dx1} and the friction force of suspension mechanism F_{dx2}, have been defined in Sect. 3.4. In turn, the human body model has been determined according to the modelling procedure presented in Sect. 3.7.2.

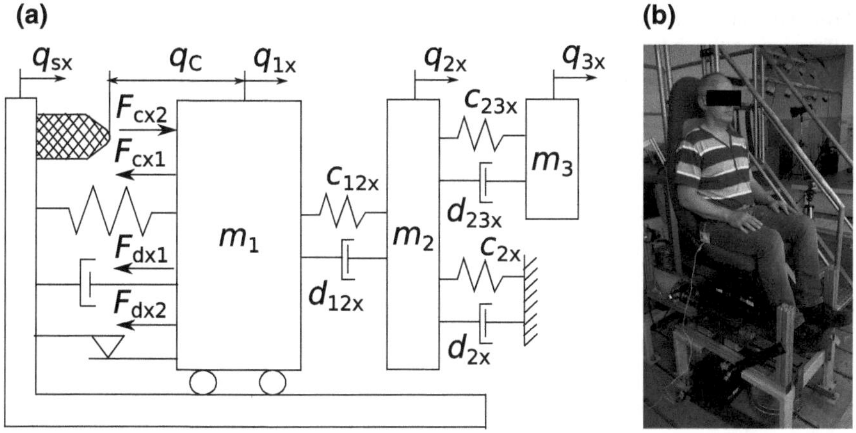

Fig. 4.10 Physical model of the horizontal seat suspension (**a**) and experimental set-up for the evaluation of its dynamic behaviour (**b**)

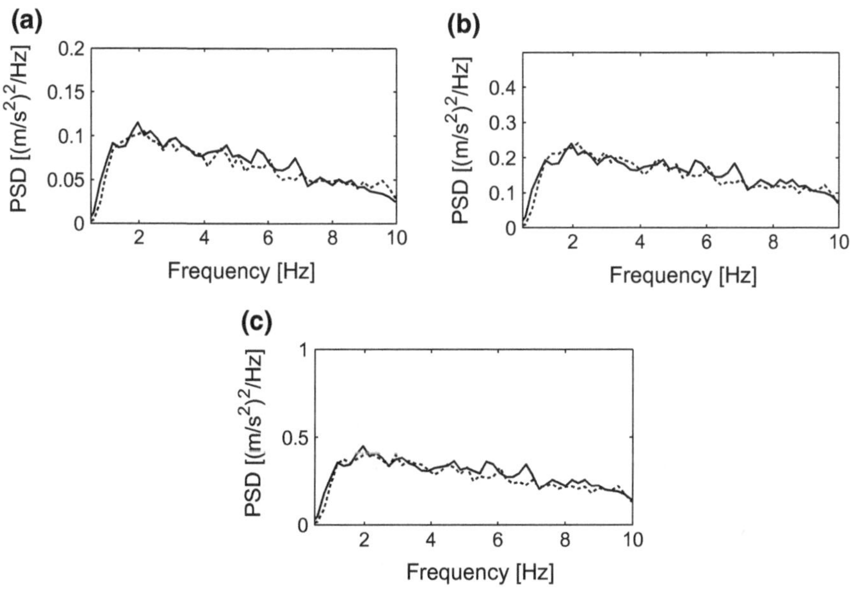

Fig. 4.11 Power spectral densities of simulated (—) and measured (----) input vibrations for the spectral classes: WN1x (**a**), WN2x (**b**) and WN3x (**c**)

During the evaluation of the operator seat vibration in horizontal direction, an excitation signal similar to the white, band-limited noise in the range of frequency 0.5–10 Hz is used. Simulated and measured power spectral densities of the input signals at different excitation intensities: 1.02 m/s^2 (WN1x), 1.36 m/s^2 (WN2x), 1.91 m/s^2 (WN3x) are presented in Fig. 4.11.

Table 4.4 Numerical values of the identified model parameters for the human body mass 90 kg at different excitation intensities

Excitation intensity, m/s^2	Mass, kg	Damping, Ns/m	Stiffness, N/m
WN1x 1.02	m_1 6.8	d_{12x} 722.6	c_{12x} 89675
	m_2 69.7	d_{23x} 413.3	c_{23x} 4253
	m_3 7.2	d_{2x} 14.5	c_{2x} 131.4
WN2x 1.36	m_1 7.6	d_{12x} 301.1	c_{12x} 50836
	m_3 70.6	d_{23x} 371.3	c_{23x} 1894
	m_3 6.7	d_{2x} 5.3	c_{2x} 149.1
WN3x 1.91	m_1 8.5	d_{12x} 233.5	c_{12x} 36551
	m_2 68.9	d_{23x} 302.3	c_{23x} 552.4
	m_3 7.1	d_{2x} 4.3	c_{2x} 99.1

Table 4.5 Simulated and measured transmissibility factors, suspension travels and relative errors of the horizontal seat suspension with seated human body obtained for excitation signals: WN1x, WN2x and WN3x, human body mass 90 kg

Input vibration	Simulation		Measurement		Relative error	
	TFE_x factor	s_{tx}, mm	TFE_x factor	s_{tx}, mm	δ_{TFE_x}, %	$\delta_{s_{tx}}$, %
WN1x	0.965	16.4	1.052	18.2	8.2	9.8
WN2x	0.965	23.2	1.051	25.1	8.2	7.5
WN3x	0.994	27.9	1.075	28.2	7.5	1.1

The male subject with body mass 90 kg is chosen for the tests, and such a subject is exposed to three vibration inputs (Fig. 4.11) in longitudinal direction. The human body model parameters (Fig. 4.10a), i.e. body masses: m_1, m_2, m_3, damping coefficients: d_{12x}, d_{23x}, d_{2x} and stiffness coefficients: c_{12x}, c_{23x}, c_{2x}, are identified for a sitting position of the human body (Fig. 4.10b). An appropriate selection of the model parameters is performed by minimising the root mean square error between the simulated and measured transmissibility functions in the frequency domain. The obtained numerical values are shown in Table 4.4.

The power spectral densities and transmissibility functions of the horizontal seat suspension with seated human body are shown in Fig. 4.12. The simulation results of the system under examination correspond to the results measured experimentally with the error not exceeding 10%. The worst agreement between vibration amplitudes can be observed in the 3–6 Hz frequency range. In this range, the cushioned seat back has a significant influence on the system dynamics. The particular transmissibility factors, suspension travels and relative errors are presented in Table 4.5.

The vibration evaluation according to this ISO-2631 [6] standard includes the measurements of the weighted root mean square (RMS) acceleration, because within the frequency range 0.1–100 Hz the vibration acceleration limits depend on vibration

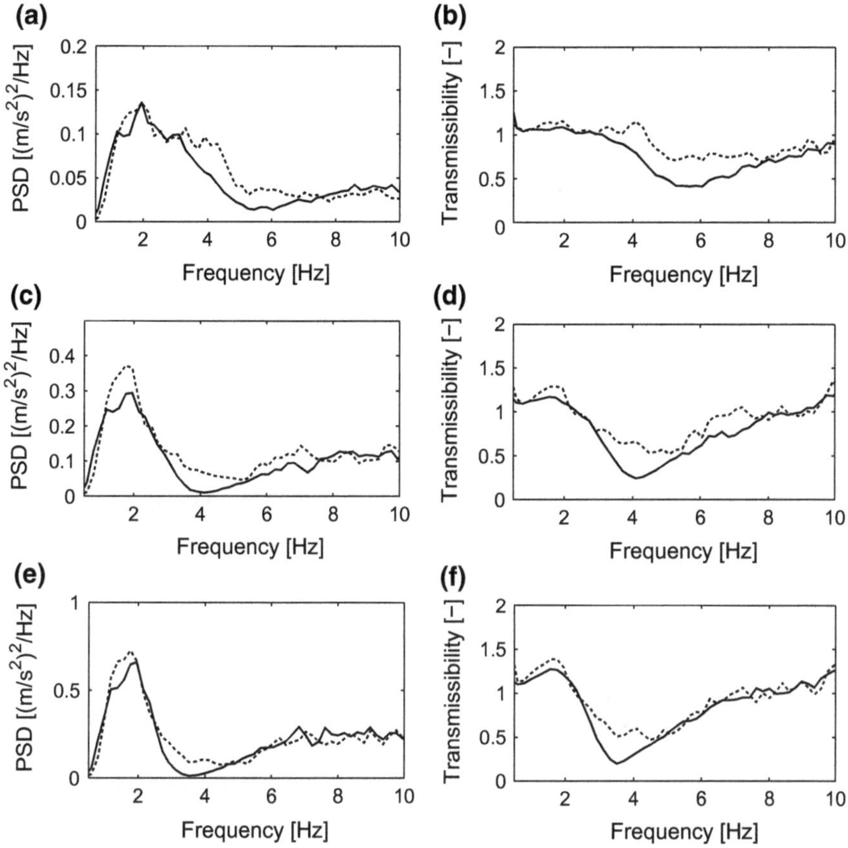

Fig. 4.12 Simulated (——) and measured (----) power spectral densities for different spectral classes: WN1x (**a**), WN2x (**c**), WN3x (**e**) and transmissibility functions for different spectral classes: WN1x (**b**), WN2x (**d**), WN3x (**f**), human body mass 90 kg

frequency. The greatest sensitivity of a seated human to horizontal acceleration is between 0.5 and 3 Hz because the human body has a number of natural frequencies in this range. Principal frequency weightings recommended for the various directions and their applications are presented in Sect. 4.3.

The simulation and measurement results obtained by using the ISO-2631 weighting method are then analysed with respect to the total daily duration of exposure to the vibration (Fig. 4.13). As shown in this figure, the root mean square (RMS) accelerations for input vibrations WN1x (Fig. 4.13a) and WN2x (Fig. 4.13c) indicate exposures in the area between the exposure action value $(a_{8h})_a = 0.5\ \mathrm{m/s^2}$ and the exposure limit value $(a_{8h})_l = 1.15\ \mathrm{m/s^2}$. These exposures cannot be considered to be safe, because may cause the vibration injury in some workers [11]. For the input vibration WN3x (Fig. 4.13e), exposure of working machine operators exceeds the

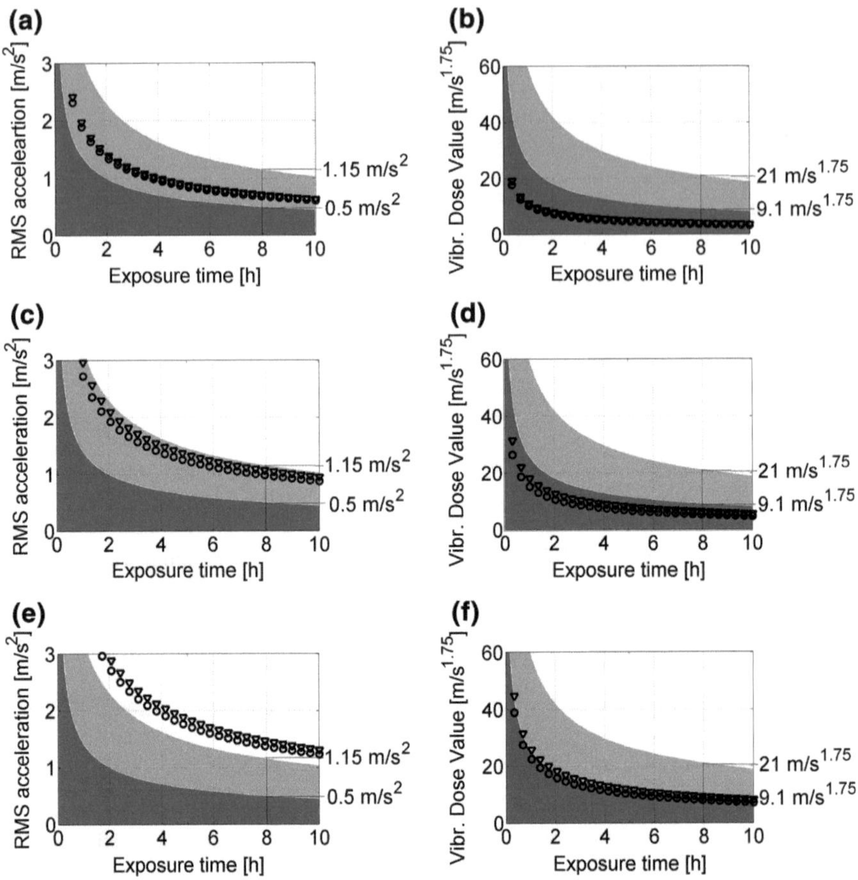

Fig. 4.13 Root mean square accelerations for different exposure to vibration input: WN1x (**a**), WN2x (**c**), WN3x (**e**) and vibration dose values for different exposure to vibration input: WN1x (**b**), WN2x (**d**), WN3x (**f**) obtained by means of computer simulations (○) and laboratory experimental tests (▽), human body mass 90 kg

exposure action value; therefore, machine operators cannot perform their work at such a high vibration intensity.

Therefore, employers are obliged to control the whole-body vibration risks of their workers, especially after many years of exposure [11]. The vibration dose values (VDV) obtained for each of the input vibrations: WN1x (Fig. 4.13b), WN2x (Fig. 4.13d) and WN3x (Fig. 4.13f) show exposures in the area below the exposure action value $(VDV_{8h})_a = 9.1\ m/s^{1.75}$; however, this cumulative value depends on both the level of vibration and the measurement duration. The simulation and measurement time for particular input vibrations have been established just on 300 s.

4.7 Conclusions

In the following chapter, human exposure to whole-body vibration is evaluated using the method defined in ISO-2631. Based on the frequency-weighting method, the human exposure to whole-body vibration is evaluated taking into account the resonances of human body parts and organs. Then the conflicted vibro-isolation criteria are proposed in order to evaluate the dynamical properties of modern vibration reduction systems. The vibro- n properties of an exemplary horizontal seat suspension are discussed, both with the dummy masses and the test person, using a laboratory method for measuring and evaluating the system effectiveness in reducing the whole-body vibration.

References

1. Griffin, M. J. (1996). *Handbook of human vibration*. London: Elsevier Academic Press.
2. British Standards Institution BS EN 13490. (2002). *Mechanical vibration—Industrial trucks—Laboratory evaluation and specification of operator seat vibration*, London.
3. International Organization for Standardization. (2003). *Agricultural wheeled tractors—Operator's seat—Laboratory measurement of transmitted vibration*, ISO 5007, Genewa.
4. International Organization for Standardization. (2000). *Earth-moving machinery—Laboratory evaluation of operator seat vibration*, ISO 7096, Genewa.
5. Engel, Z. (1993). *Environmental protection against vibration and sound* (in Polish). Warsaw: PWN.
6. International Organization for Standardization. (1997). *Mechanical vibration and shock - Evolution of human exposure to whole body vibration*, ISO 2631, Genewa.
7. Hostens, I., & Ramon, H. (2003). Descriptive analysis of combine cabin vibrations and their effect on the human body. *Journal of Sound and Vibration*, *266*, 453–464.
8. Nader, M. (2001). *Modelling and simulation of vehicle vibration influence on human body (in Polish)*. Warsaw: Publishing House of Warsaw University of Technology.
9. Rimmel, A. N., & Mansfield, N. J. (2007). Design of digital filters for frequency weightings reguired for risk assessment of workers exposed to vibration. *Industrial Health*, *45*, 512–519.
10. Griffin, M. J., Howarth, H. V. C., Pitts, P. M., Fischer, S., Kaulbars, U., Donati, P. M., et al. (2006). *Guide to good practice on whole-body vibration. Non-binding guide to good practice with a view to implementation of Directive 2002/44/EC on the minimum health and safety requirements regarding the exposure of workers to the risks arising from physical agents (vibrations)*. Luxembourg: European Commission.
11. Directive 2002/44/EC of the European Parliament and of the Council. (2002). *On the minimum health and safety requirements regarding the exposure of workers to the risks arising from physical agents (vibration)*, Official Journal of the European Communities, pp. 13–18.
12. Maciejewski, I., Meyer, L., & Krzyzynski, T. (2009). Modelling and multi-criteria optimisation of passive seat suspension vibro-isolating properties. *Journal of Sound and Vibration*, *324*, 520–538.
13. Paddan, G., & Griffi, M. (2001). *Use of seating to control exposures to whole-body vibration*. London: Health and Safety Executive.

Chapter 5
Multi-criteria Optimisation of the Vibro-isolation Properties

5.1 Nomenclature

$\mathbf{F}(\mathbf{x}_d)$	vector of the optimisation criteria
$\mathbf{S}_{ti}$	vector of the suspension travel for selected direction of the vibration exposure ($i = x, y, z$)
S_j	first-order sensitivity index
S_{Tj}	total effect sensitivity index
$TFE_i(\mathbf{x}_d)$	frequency-weighted transmissibility factor for selected direction of the vibration exposure ($i = x, y, z$)
$\mathbf{X}_A, \mathbf{X}_B, \mathbf{X}_{Cj}$	sample matrices
$f_i(\mathbf{x}_d)$	optimisation criteria ($i \geq 2$)
$g_i(\mathbf{x}_d)$	inequality and equality constraints ($i = 1, \ldots, m$)
$s_{ti}(\mathbf{x}_d)$	suspension travel for selected direction of the vibration exposure ($i = x, y, z$), m
$\mathbf{x}_d$	vector of the selected decision variables
$(\mathbf{x}_d)_{max}$	vector of the maximal values of design parameters
$(\mathbf{x}_d)_{min}$	vector of the minimal values of design parameters
ϵ_i	constraint value of the ith criterion

5.2 Introduction

A design process of the vibration reduction systems is complicated because of their opposite performance requirements [1]. For example, in typical working machines, it is desirable to reduce the vibration transmission from its source to the human operator. Thus, the vibration isolation of operators can be achieved by applying the seat suspension system [2]. From the first point of view, the vibrations transmitted into the body of the operator should be minimised in order to meet health and safety requirements during the workday. Against this background, the suspension relative

T. Krzyzynski et al., *Modelling and Control Design of Vibration Reduction Systems*, https://doi.org/10.1007/978-3-030-03047-6_5

motion should also be minimised in order to provide the desired controllability of working machine by the operator. The multi-criteria optimisation can be used to find the compromise solution with respect to the conflicted vibro-isolation criteria.

5.3 Problem Formulation

The primary goal of multi-criteria optimisation is to discover the compromise solutions in view of several opposing criteria. Such solutions provide optimal system performance for the conflicted requirements, i.e. system effectiveness, performance robustness, energy savings, etc. In this case, a vector of the criteria represents the problem has to be defined as:

$$\mathbf{F}(\mathbf{x}_\mathrm{d}) = [f_1(\mathbf{x}_\mathrm{d}), f_2(\mathbf{x}_\mathrm{d}), \ldots, f_\mathrm{i}(\mathbf{x}_\mathrm{d})]^\mathrm{T}, \qquad \mathrm{i} \geq 2 \tag{5.1}$$

where $f_1, f_2, \ldots, f_\mathrm{i}$ are the particular optimisation criteria and $\mathbf{x}_\mathrm{d}$ are the decision variables. Multi-criteria optimisation is concerned with the minimisation of a vector of criteria $\mathbf{F}(\mathbf{x}_\mathrm{d})$ that can be the subject of a number of constraints or bounds as follows:

$$\min_{\mathbf{x}_\mathrm{d}} \mathbf{F}(\mathbf{x}_\mathrm{d}) \tag{5.2}$$

subject to the following constraints:

- inequality constraints

$$g_\mathrm{i}(\mathbf{x}_\mathrm{d}) \leq 0, \qquad \mathrm{i} = 1, \ldots, \mathrm{n} \tag{5.3}$$

- equality constraints

$$g_\mathrm{i}(\mathbf{x}_\mathrm{d}) = 0, \qquad \mathrm{i} = \mathrm{n} + 1, \ldots, \mathrm{m} \tag{5.4}$$

and the following ranges of decision variables:

$$(\mathbf{x}_\mathrm{d})_\mathrm{min} \leq \mathbf{x}_\mathrm{d} \leq (\mathbf{x}_\mathrm{d})_\mathrm{max} \tag{5.5}$$

where $(\mathbf{x}_\mathrm{d})_\mathrm{min}$ and $(\mathbf{x}_\mathrm{d})_\mathrm{max}$ are the vectors of the minimal and maximal values of selected decision variables.

In order to solve this problem, the concept of Pareto optimality should be used. Non-dominated solutions (synonyms are: Pareto-optimal solutions, non-inferior solutions) have to be found in the feasible region of solutions. These non-dominated solutions are characterised by the fact that any attempt to improve the selected criterion must be associated with the deterioration of other optimisation criteria [3]. In

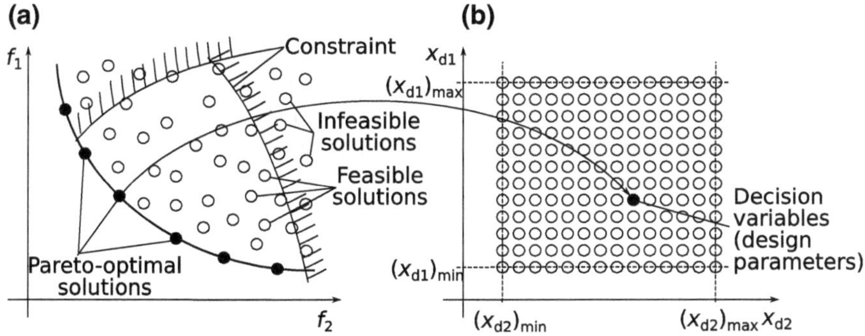

Fig. 5.1 Graphical illustration of the multi-criteria optimisation procedure: criterion space (**a**) and decision variable space (**b**)

Fig. 5.1, the method of searching non-dominated solutions by means of multi-criteria optimisation is graphically illustrated for a two-dimensional problem.

The paper [4] presents a survey of concepts and methods for the multi-criteria optimisation with an eye towards engineering applications. Each of the methods is concerned with the generation and selection of non-dominated solutions. The convexity problem arises in some methods when the Pareto-front is non-convex [5]. In this case, the set of non-dominated solutions in the non-convex region is sometimes not available (e.g. weighted sum method). A procedure that overcomes some of the convexity problems is the constraint method [6, 7]. This involves minimising a primary criterion $f_1(\mathbf{x}_d)$ as follows:

$$\min_{\mathbf{x}_d} f_1(\mathbf{x}_d) \tag{5.6}$$

and expressing the other criteria in the form of inequality constraints:

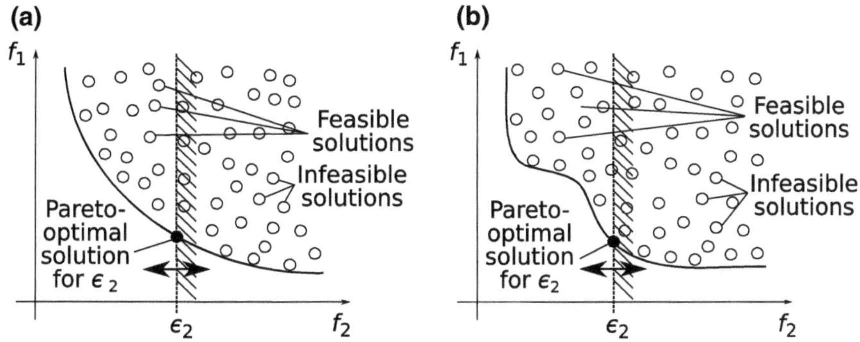

Fig. 5.2 Two-dimensional representation of the constraint method: convex Pareto-front (**a**) and non-convex Pareto-front (**b**)

$$f_i(\mathbf{x}_d) \leq \epsilon_i, \qquad i = 2, \ldots, m \tag{5.7}$$

where ϵ_i is the constraint value of the ith criterion. A problem with this method is that the suitable selection of ϵ_i ensures a feasible solution. Graphical representation of the constraint method for a bi-criteria problem is shown in Fig. 5.2.

5.4 Global Sensitivity Analysis

Global sensitivity analysis is a useful tool supporting assessment of the variations in the model outputs relatively to variations imposed on the model inputs [8]. This kind of analysis may be used to explore an influence of the model parameters on the system performance and further for determining the dominant controls of modelled system. The procedure of evaluating the relative impact of input factors on system evaluation criteria is discussed complementary in the following section.

5.4.1 Sample Matrix

The sample matrix $\mathbf{X}_{AB}$ must be created in the presence of the probability distributions that are imposed on the selected input factors. This can be done by using an uniform random number generator (rand); therefore, the sensitivity analysis is initiated accepting the following starting points:

$$\mathbf{X}_{AB} = \text{rand}(N, 2p) \tag{5.8}$$

where N describes the number of the required simulation runs (number of rows in the sample matrix). The number of columns is the doubled number 2p of input factors that would be investigated during the sensitivity analysis. The resulting sample matrix is then occurred as:

$$\mathbf{X}_{AB} = \underbrace{\begin{bmatrix} x_{A11} & x_{A12} & \cdots & x_{A1j} \\ x_{A21} & x_{A22} & \cdots & x_{A2j} \\ \cdots & \cdots & \ddots & \cdots \\ x_{Ai1} & x_{Ai2} & \cdots & x_{Aij} \end{bmatrix}}_{\mathbf{X}_A} \underbrace{\begin{bmatrix} x_{B11} & x_{B12} & \cdots & x_{B1j} \\ x_{B21} & x_{B22} & \cdots & x_{B2j} \\ \cdots & \cdots & \ddots & \cdots \\ x_{Bi1} & x_{Bi2} & \cdots & x_{Bij} \end{bmatrix}}_{\mathbf{X}_B}, i = 1, \ldots, p, \quad j = 1, \ldots, N \tag{5.9}$$

where $\mathbf{X}_A$ and $\mathbf{X}_B$ are the independent sample matrices that include two different steps of starting points in the p-dimensional unit hypercube.

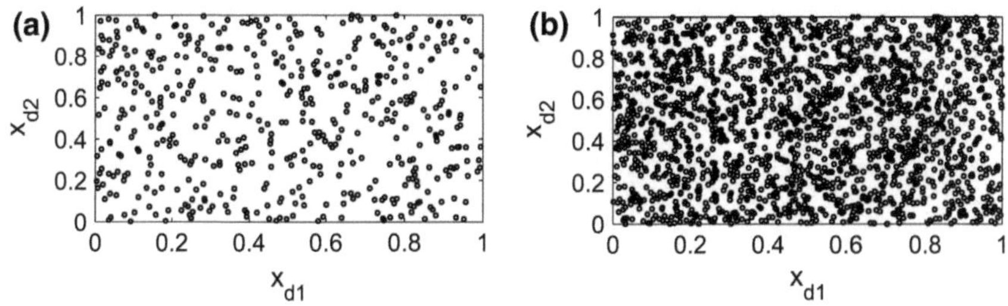

Fig. 5.3 Two-dimensional representation of uniformly distributed random starting points at a different number of simulation runs: $N = 500$ (**a**) and $N = 2000$ (**b**)

The number of N samples influences the quality of sensitivity indices to be estimated [9]. If the number of N has higher values, then the better estimation of sensitivity indices can be achieved. Exemplary representation of the uniformly distributed random starting points at two specific values of N samples is shown in Fig. 5.3.

5.4.2 Variance-Based Method

Typically, the sensitivity analysis is performed by measuring only two levels of input factors, i.e. at the lower and upper bounds [10]. In contrast to the conventional design process, the variance-based method is based on analysing the whole space of input factors employing the Monte Carlo method [11]. Hence, the variance-based method takes into account the non-linear effects of particular input factors and lets recognise the interactive effects in non-additive systems [12].

The random starting points are exploited for the purpose of calculating the output variances. The model outputs obtained in this way must be decomposed into relevant fractions, and their values shall be attributed to the system inputs [13]. Such a method uses the combination of two independent sampling matrices $\mathbf{X}_{\mathrm{A}}$ and $\mathbf{X}_{\mathrm{B}}$ (Eq. (5.9)) in order to compute the recombined matrices $\mathbf{X}_{\mathrm{C1}}, \mathbf{X}_{\mathrm{C2}}, \ldots, \mathbf{X}_{\mathrm{Cj}}$. It should be noted that the output matrices contain the elements of matrix $\mathbf{X}_{\mathrm{A}}$ except the jth column which is taken from matrix $\mathbf{X}_{\mathrm{B}}$ as shown below:

$$\mathbf{X}_{\mathrm{C1}} = \begin{bmatrix} x_{\mathrm{B11}} & x_{\mathrm{A12}} & \cdots & x_{\mathrm{A1j}} \\ x_{\mathrm{B21}} & x_{\mathrm{A22}} & \cdots & x_{\mathrm{A2j}} \\ \cdots & \cdots & \ddots & \cdots \\ x_{\mathrm{Bi1}} & x_{\mathrm{Ai2}} & \cdots & x_{\mathrm{Aij}} \end{bmatrix}, \quad \mathrm{i} = 1, \ldots, N, \quad \mathrm{j} = 1, \ldots, \mathrm{p}$$

$$\mathbf{X}_{\mathrm{C2}} = \begin{bmatrix} x_{\mathrm{A11}} & x_{\mathrm{B12}} & \cdots & x_{\mathrm{A1j}} \\ x_{\mathrm{A21}} & x_{\mathrm{B22}} & \cdots & x_{\mathrm{A2j}} \\ \cdots & \cdots & \ddots & \cdots \\ x_{\mathrm{Ai1}} & x_{\mathrm{Bi2}} & \cdots & x_{\mathrm{Aij}} \end{bmatrix}, \quad \mathrm{i} = 1, \ldots, N, \quad \mathrm{j} = 1, \ldots, \mathrm{p} \tag{5.10}$$

$$\vdots$$

$$\mathbf{X}_{\mathrm{Cj}} = \begin{bmatrix} x_{\mathrm{A11}} & x_{\mathrm{A12}} & \cdots & x_{\mathrm{B1j}} \\ x_{\mathrm{A21}} & x_{\mathrm{A22}} & \cdots & x_{\mathrm{B2j}} \\ \cdots & \cdots & \ddots & \cdots \\ x_{\mathrm{Ai1}} & x_{\mathrm{Ai2}} & \cdots & x_{\mathrm{Bij}} \end{bmatrix}, \quad \mathrm{i} = 1, \ldots, N, \quad \mathrm{j} = 1, \ldots, \mathrm{p}$$

The model outputs have to be computed for each design point in the $\mathbf{X}_{\mathrm{A}}$, $\mathbf{X}_{\mathrm{B}}$ and $\mathbf{X}_{\mathrm{C1}}, \mathbf{X}_{\mathrm{C2}}, \ldots, \mathbf{X}_{\mathrm{Cj}}$ matrices; thus, a total of $N(2 + \mathrm{p})$ model calculations are recommended. The equivalent model outputs would be reported as the following vectors:

$$f(\mathbf{X}_{\mathrm{A}}) = \begin{bmatrix} f\,(\mathbf{X}_{\mathrm{A}})_1 \\ f\,(\mathbf{X}_{\mathrm{A}})_2 \\ \cdots \\ f\,(\mathbf{X}_{\mathrm{A}})_{\mathrm{i}} \end{bmatrix}, f(\mathbf{X}_{\mathrm{B}}) = \begin{bmatrix} f\,(\mathbf{X}_{\mathrm{B}})_1 \\ f\,(\mathbf{X}_{\mathrm{B}})_2 \\ \cdots \\ f\,(\mathbf{X}_{\mathrm{B}})_{\mathrm{i}} \end{bmatrix}, \quad \begin{array}{c} \mathrm{i} = 1, \ldots, N \\ \mathrm{j} = 1, \ldots, \mathrm{p} \end{array} \tag{5.11}$$

$$f(\mathbf{X}_{\mathrm{C1}}) = \begin{bmatrix} f\,(\mathbf{X}_{\mathrm{C1}})_1 \\ f\,(\mathbf{X}_{\mathrm{C1}})_2 \\ \cdots \\ f\,(\mathbf{X}_{\mathrm{C1}})_{\mathrm{i}} \end{bmatrix}, f(\mathbf{X}_{\mathrm{C2}}) = \begin{bmatrix} f\,(\mathbf{X}_{\mathrm{C2}})_1 \\ f\,(\mathbf{X}_{\mathrm{C2}})_2 \\ \cdots \\ f\,(\mathbf{X}_{\mathrm{C2}})_{\mathrm{i}} \end{bmatrix}, f(\mathbf{X}_{\mathrm{Cj}}) = \begin{bmatrix} f\,(\mathbf{X}_{\mathrm{Cj}})_1 \\ f\,(\mathbf{X}_{\mathrm{Cj}})_2 \\ \cdots \\ f\,(\mathbf{X}_{\mathrm{Cj}})_{\mathrm{i}} \end{bmatrix}$$

These vectors $f(\mathbf{X}_{\mathrm{A}})$; $f(\mathbf{X}_{\mathrm{B}})$; $f(\mathbf{X}_{\mathrm{C1}}), f(\mathbf{X}_{\mathrm{C2}}), \ldots, f(\mathbf{X}_{\mathrm{Cj}})$ consist of numerically evaluated assessment criteria that are achieved for the ith set of initial starting points.

5.4.3 *Sensitivity Indices*

Sensitivity analysis is curried out based on of the first-order sensitivity index S_{j} and the total effect index S_{Tj}. The first-order sensitivity index (S_{j}) demonstrates essential effects of the model output variance regarding the jth input factor. Hence, the total effect sensitivity index (S_{Tj}) consists of the higher-order effects as the sum of the first, second, third and more complex effects [9]. As follows from the paper [11], the first-order sensitivity index S_{j} and the total effect sensitivity index S_{Tj} may be computed by introducing the subsequent relations:

$$S_{\mathrm{j}} = 1 - \frac{\frac{1}{2N}\sum_{\mathrm{i}=1}^{N}\left(f(\mathbf{X}_{\mathrm{B}})_{\mathrm{i}} - f(\mathbf{X}_{\mathrm{Cj}})_{\mathrm{i}}\right)^2}{\frac{1}{N}\sum_{\mathrm{i}=1}^{N} f^2(\mathbf{X}_{\mathrm{A}})_{\mathrm{i}} - f_0^2}, \quad \mathrm{i} = 1, \ldots, N, \ \mathrm{j} = 1, \ldots, \mathrm{p} \tag{5.12}$$

$$S_{\mathrm{Tj}} = \frac{\frac{1}{2N}\sum_{\mathrm{i}=1}^{N}\left(f(\mathbf{X}_{\mathrm{A}})_{\mathrm{i}} - f(\mathbf{X}_{\mathrm{Cj}})_{\mathrm{i}}\right)^2}{\frac{1}{N}\sum_{\mathrm{i}=1}^{N} f^2(\mathbf{X}_{\mathrm{A}})_{\mathrm{i}} - f_0^2}, \quad \mathrm{i} = 1, \ldots, N, \ \mathrm{j} = 1, \ldots, \mathrm{p} \tag{5.13}$$

where $f(\mathbf{X}_{\mathrm{B}})_{\mathrm{i}}$ and $f(\mathbf{X}_{\mathrm{A}})_{\mathrm{i}}$ are the model outputs determined adopting independent sets of initial starting points and $f(\mathbf{X}_{\mathrm{Cj}})_{\mathrm{i}}$ is again the model output; however, it is computed by using the recombined sets of initial starting points (see previous Sect. 5.4.2). Expected value f_0 of the model outputs shall be calculated by means of arithmetic mean as follows [11]:

$$f_0 = \frac{1}{N}\sum_{\mathrm{i}=1}^{N} f(\mathbf{X}_{\mathrm{A}})_{\mathrm{i}},, \quad \mathrm{i} = 1, \ldots, N \tag{5.14}$$

where N is the number of simulation runs.

5.5 Example: Sensitivity Analysis of the Horizontal Seat Suspension

The sensitivity analysis of the horizontal seat suspension is performed by using simulation model of the system (see Sect. 3.5). For the purpose of adapting its force characteristics, the following input factors and their variability ranges are taken as:

- wire diameter of the helical springs $d_{\mathrm{s}} = 1.4 - 2.2 \cdot 10^{-3}$ m,
- linear stiffness coefficient of the end-stop buffers $c_{\mathrm{b1}} = 3 - 30 \cdot 10^3$ N/m,
- cubic stiffness coefficient of the end-stop buffers $c_{\mathrm{b3}} = 50 - 500 \cdot 10^6$ N/m^3,
- orifice diameter of the shock-absorber $d_{\mathrm{o}} = 0.75 - 1.15 \cdot 10^{-3}$ m,
- orifice length of the shock-absorber $l_{\mathrm{o}} = 2 - 20 \cdot 10^{-3}$ m,
- reduction ratio of the friction force $\delta_{\mathrm{f}} = 0.2 - 1.2$.

The model outputs are defined as the set of vibro-isolation criteria, i.e. the frequency-weighted transmissibility factor $\mathrm{TFE}_{\mathrm{x}}$ and the suspension travel s_{tx} that are calculated for x-axis of the vibration transmission. Random number generator (rand) is used to generate two independent sets of initial starting points that is matrix X_{A} and X_{B} corresponding to the model outputs $f(\mathbf{X}_{\mathrm{A}})_{\mathrm{i}}$, $f(\mathbf{X}_{\mathrm{B}})_{\mathrm{i}}$. They are calculated twice for each of the evaluation criteria. The matrices $f(\mathbf{X}_{\mathrm{Cj}})_{\mathrm{i}}$ also include the vibro-isolation criteria; however, they are obtained for the recombined sample matrices $\mathbf{X}_{\mathrm{Cj}}$ (see Sect. 5.4.2). In this study, the number of samples is set at $N = 1000$ points and the number of input factors is $\mathrm{p} = 6$; therefore, a total number of simulation runs are equal to $N(2 + \mathrm{p}) = 8000$. First-order and total effect sensitivity indices of the transmissibility factor $\mathrm{TFE}_{\mathrm{x}}$ and suspension travel s_{tx} are evaluated for different

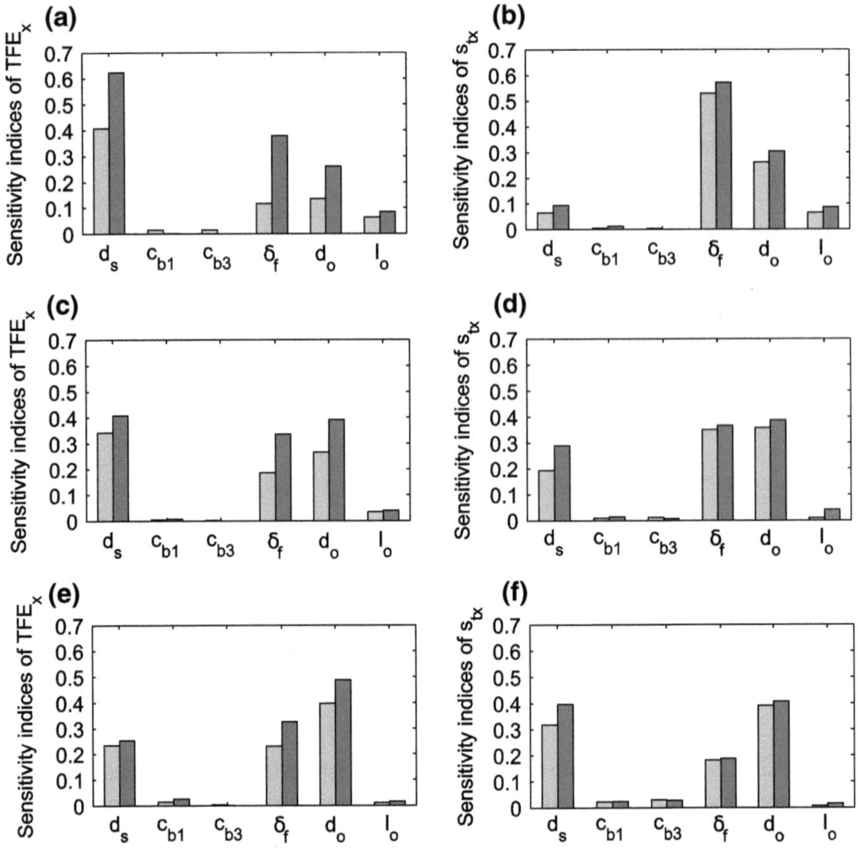

Fig. 5.4 First-order ( ) and total effect ( ) sensitivity indices of transmissibility factor $\mathrm{TFE_x}$ for different input vibrations: WN1x (**a**), WN2x (**c**), WN3x (**e**) and sensitivity indices of suspension travel s_{tx} for different input vibrations: WN1x (**b**), WN2x (**d**), WN3x (**f**)

input vibrations, i.e. WN1x, WN2x, WN3x (see Sect. 4.6). Graphical representation of the obtained results is presented in Fig. 5.4.

As shown in Fig. 5.4, the lowest sensitivities on model outputs are obtained for the linear c_{b1} and cubic c_{b3} stiffness coefficients of end-stop buffers. The orifice length l_{o} of shock-absorber also points out the low sensitivity on the system evaluation criteria; however, the higher sensitivity is achieved at the lowest excitation intensity (signal WN1x). The other input factors, i.e. the wire diameter d_{s} of helical springs, the orifice diameter d_{o} of shock-absorber and the reduction ratio δ_{f} of friction force, influence significantly the first-order and total effect sensitivity indices of the transmissibility factor $\mathrm{TFE_x}$ and suspension travel s_{tx}. Therefore, further analysis of the system dynamics is performed on the basis of selected input factors: d_{s}, d_{o} l_{o} and δ_{f}.

5.6 Optimisation Procedure

In the following section, the compromised solutions (Pareto-optimal solutions) are searched using the proposed optimisation procedure. At first, the conflicted vibro-isolation criteria are minimised separately. Then, the primary optimisation criterion is defined subject to the other evaluation criteria that are transferred to the non-linear inequality constraint. The decision variables are chosen as the design parameters allowing to change the dynamic characteristics of vibration reduction system.

5.6.1 Decision Variables

The following vector should represent the chosen decision variables:

$$\mathbf{x}_{\mathrm{d}} = [x_{\mathrm{d1}}, x_{\mathrm{d2}}, \ldots, x_{\mathrm{dn}}]^{\mathrm{T}}, \qquad \mathrm{n} = 1, \ldots, \mathrm{r} \tag{5.15}$$

where p is the total number of decision variables. Their variability ranges are defined by the following vectors:

- lower limit

$$(\mathbf{x}_{\mathrm{d}})_{\mathrm{min}} = [(x_{\mathrm{d1}})_{\mathrm{min}}, (x_{\mathrm{d2}})_{\mathrm{min}}, \ldots, (x_{\mathrm{dn}})_{\mathrm{min}}]^{\mathrm{T}} \tag{5.16}$$

- upper limit

$$(\mathbf{x}_{\mathrm{d}})_{\mathrm{max}} = [(x_{\mathrm{d1}})_{\mathrm{max}}, (x_{\mathrm{d2}})_{\mathrm{max}}, \ldots, (x_{\mathrm{dn}})_{\mathrm{max}}]^{\mathrm{T}} \tag{5.17}$$

where $(x_{\mathrm{d1}})_{\mathrm{min}}, (x_{\mathrm{d2}})_{\mathrm{min}}, \ldots, (x_{\mathrm{dn}})_{\mathrm{min}}$ are the minimum values of decision variables and $(x_{\mathrm{d1}})_{\mathrm{max}}, (x_{\mathrm{d2}})_{\mathrm{max}}, \ldots, (x_{\mathrm{dn}})_{\mathrm{max}}$ are the maximum values of decision variables.

5.6.2 Random Starting Points

Most often, the optimisation algorithms attempt to find a minimum of the objective function starting at an initial estimate. The initial starting points of decision variables have a significant influence on the found optimum, especially in the non-linear tasks (problem of local minima). For most non-linear problems, this initial estimate issue using random staring points is recommended [14]. Such procedure ensures a high probability that the optimum found is a global one.

Using a uniform random number generator, the initial starting points for optimisation procedure may be expressed as follows:

$$\begin{aligned} (x_{d1})_s &= (x_{d1})_{min} + ((x_{d1})_{max} - (x_{d1})_{min})\,\mathrm{rand(j)} \\ (x_{d2})_s &= (x_{d2})_{min} + ((x_{d2})_{max} - (x_{d2})_{min})\,\mathrm{rand(j)} \\ \ldots & \qquad\qquad\qquad\qquad\qquad\qquad \mathrm{j} = 1, \ldots, \mathrm{k} \\ (x_{dr})_s &= (x_{dr})_{min} + ((x_{dr})_{max} - (x_{dr})_{min})\,\mathrm{rand(j)} \end{aligned} \tag{5.18}$$

where i is the number of pseudo-random values (rand) drawn from the standard uniform distribution on the closed interval [0,1].

5.6.3 Minimisation of Particular Optimisation Criteria

The separate minimising of the transmissibility factor ($\mathrm{TFE_i}$) and the suspension travel (s_{ti}) is defined masses as follows:

$$\min_{\mathbf{x}_d} \ \mathrm{TFE_i}(\mathbf{x}_d), \qquad \mathrm{i} = x, y, z \tag{5.19}$$

$$\min_{\mathbf{x}_d} \ s_{ti}(\mathbf{x}_d), \qquad \mathrm{i} = x, y, z \tag{5.20}$$

subject to the following ranges of decision variables:

$$(\mathbf{x}_d)_{min} \le \mathbf{x}_d \le (\mathbf{x}_d)_{max} \tag{5.21}$$

where $(\mathbf{x}_d)_{min}$ and $(\mathbf{x}_d)_{max}$ are the vectors describing minimum and maximum values of the chosen decision variables, respectively.

Such an optimisation procedure allows to find the marginal solutions which minimises the particular optimisation criteria separately:

- criterion coordinates $(s_{ti})_{max}$, $(\mathrm{TFE_i})_{min}$—minimum of the transmissibility factor,
- criterion coordinates $(s_{ti})_{min}$, $(\mathrm{TFE_i})_{max}$—minimum of the suspension travel.

The graphical representation of minimising the particular optimisation criteria is shown in Fig. 5.5. The optimised systems are characterised by the best reduction of forces transmitted to the isolated body (soft suspension system) or by the best limitation of the suspension travel (stiff suspension system).

5.6.4 Suspension Travel Range

The linearly spaced vector of suspension travel is defined as follows:

$$\mathbf{S}_{ti} = [s_{ti1}, s_{ti2}, \ldots, s_{tij}], \qquad \mathrm{i} = x, y, z, \qquad \mathrm{j} = 1, \ldots, \mathrm{u} \tag{5.22}$$

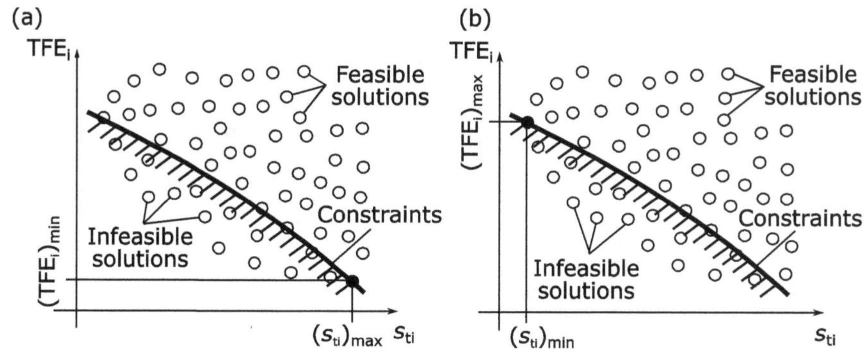

Fig. 5.5 Graphical representation of minimising the particular optimisation criteria: transmissibility factor (**a**) and suspension travel (**b**)

where u is the number of requested solutions. The vector elements are described by the following expressions:

$$\begin{aligned}
s_{\mathrm{ti1}} &= (s_{\mathrm{ti}})_{\min} + \tfrac{1-1}{\mathrm{u}-1}\left((s_{\mathrm{ti}})_{\max} - (s_{\mathrm{ti}})_{\min}\right)\\
s_{\mathrm{ti2}} &= (s_{\mathrm{ti}})_{\min} + \tfrac{2-1}{\mathrm{u}-1}\left((s_{\mathrm{ti}})_{\max} - (s_{\mathrm{ti}})_{\min}\right)\\
&\ldots\\
s_{\mathrm{tij}} &= (s_{\mathrm{ti}})_{\min} + \tfrac{\mathrm{j}-1}{\mathrm{u}-1}\left((s_{\mathrm{ti}})_{\max} - (s_{\mathrm{ti}})_{\min}\right)
\end{aligned} \tag{5.23}$$

where u is the number of points linearly spaced between the minimum $(s_{\mathrm{ti}})_{\min}$ and the maximum $(s_{\mathrm{ti}})_{\max}$ values of suspension travel.

5.6.5 *Minimising Both of the Conflicted Vibro-isolation Criteria*

In order to optimise both of the conflicted vibro-isolation criteria, a minimising of the $\mathrm{TFE_i}$ factor (primary criterion) is defined as follows:

$$\min_{\mathbf{x}_{\mathrm{d}}} \ \mathrm{TFE_i}(\mathbf{x}_{\mathrm{d}}), \qquad s_{\mathrm{tij}} \in \mathbf{S}_{\mathrm{ti}}, \qquad \mathrm{i} = x, y, z, \qquad \mathrm{j} = 1, \ldots, \mathrm{u} \tag{5.24}$$

subject to the suspension travel s_{ti} that is transferred to a non-linear inequality constraint:

$$s_{\mathrm{ti}}(\mathbf{x}_{\mathrm{d}}) \le s_{\mathrm{tij}}, \qquad \mathrm{i} = x, y, z, \qquad \mathrm{j} = 1, \ldots, \mathrm{u} \tag{5.25}$$

and the following ranges of particular decision variables:

$$(\mathbf{x}_{\mathrm{d}})_{\min} \le \mathbf{x}_{\mathrm{d}} \le (\mathbf{x}_{\mathrm{d}})_{\max} \tag{5.26}$$

where s_{tij} is the constraint value of suspension travel.

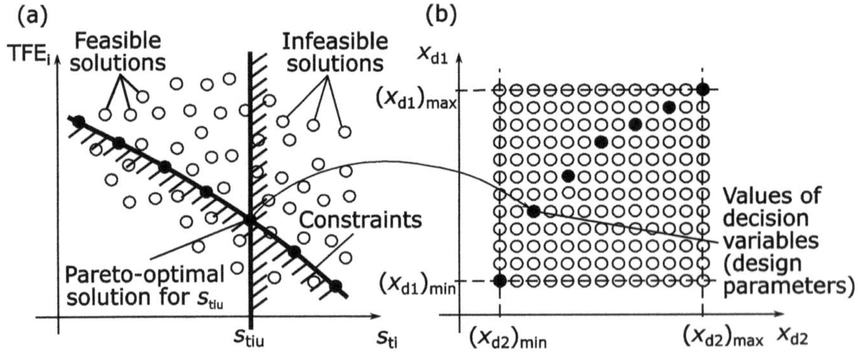

Fig. 5.6 Graphical representation of minimising both of the conflicted optimisation criteria: criterion space (**a**) and decision variable space (**b**)

The graphical representation of minimising both of the conflicted vibro-isolation criteria is shown in Fig. 5.6. This figure shows that an appropriate selection of the constraint value s_{tij} allows to choose the vibro-isolation properties of vibration reduction system. A very stiff suspension (the marginal solution on the left-hand side) can be transformed to a very soft one (the marginal solution on the right-hand side) by an appropriate selection of the system characteristics that are defined by the values of chosen decision variables.

5.7 Example: Multi-criteria Optimisation of the Horizontal Seat Suspension

The non-linear shaping of the system characteristics is required to improve vibration isolation of the horizontal seat suspension. For this reason, a modification of the following visco-elastic elements is considered: helical springs by means of the wire diameter d_s, hydraulic shock-absorber by using the orifice geometry (diameter d_o and length l_o) and overall friction of the suspension mechanism employing the reduction ratio δ_f. These model parameters are selected as the decision variables that essentially influence the system performance. The frequency-weighted transmissibility factor TFE_x and the suspension travel s_{tx} are adopted as the conflicted vibro-isolation criteria (see Sect. 4.4). The TFE_x factor is accepted as a fundamental indicator of the vibration reduction efficiency. Then, the suspension travel s_{tx} is assigned as the non-linear inequality constraint during optimisation procedure (see Sect. 5.6). Such a configuration of the optimisation problem makes possible to find relatively low values of the transmissibility factor for previously defined suspension travel. The optimisation procedure is realised for randomly generated starting points: firstly, the particular criteria are minimised, and secondly, both of the conflicted vibro-isolation criteria are optimised.

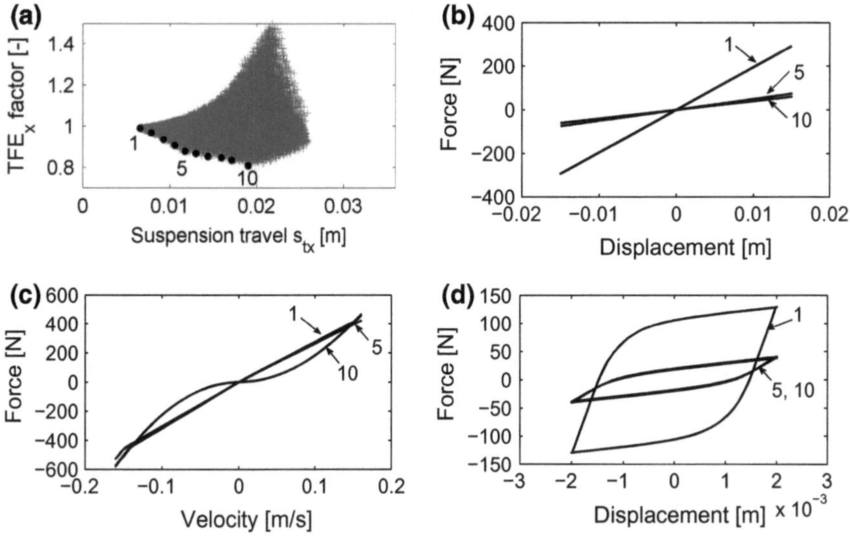

Fig. 5.7 Pareto-optimal point distribution for the input vibration WN1x (**a**) and identified visco-elastic characteristics of the helical spring (**b**), hydraulic shock-absorber (**c**) and friction of the suspension mechanism (**d**)

In Figs. 5.7, 5.8 and 5.9, the optimisation results are shown. These results are obtained for the excitation signal similar to white, band-limited noise in the range of frequency 0.5-10 Hz at different excitation intensities, i.e. WN1x, WN2x and WN3x (see Sect. 4.6). Such a choice of the excitation signals allows to investigate the seat suspension dynamic behaviour for different working conditions. However, the other excitation signals are also applicable to the proposed seat suspension design and could be considered similarly.

In the presented example, ten Pareto-optimal solutions for each excitation signal are found. Each individual Pareto-optimal point corresponds to a set of the decision variables that defines different vibro-isolation characteristics of the passive suspension system. The extreme Pareto-optimal points that are located on the left-hand side of each figure (Figs. 5.7a, 5.8a and 5.9a) correspond to the stiff suspension system (high limitation of the suspension travel). Unfortunately, a reduction of the TFE_x factor is rather poor in this case. The extreme Pareto-optimal points presented on the right-hand side correspond to the soft suspension system (low limitation of the suspension travel), but the transmissibility factor has decreased significantly. The compromising solutions of suspension system are located in between the extreme points. A selection one of the Pareto-optimal system should be made by the machine operators whose preferences and satisfaction levels have to be taken into account.

The control of the vibro-isolation properties of the seat suspension is enabled by changing the values of the decision variables. Individual Pareto-optimal solutions correspond to the set of decision variables which in turn represent visco-elastic

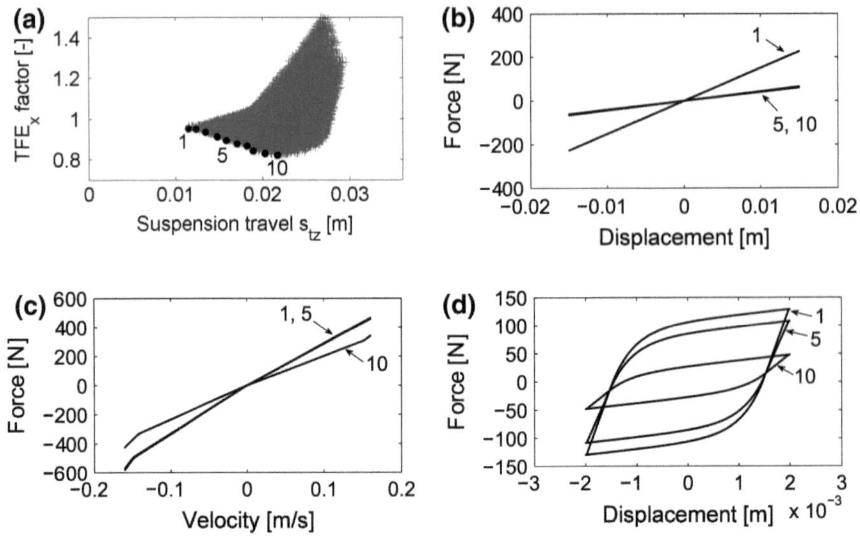

Fig. 5.8 Pareto-optimal point distribution for the input vibration WN2x (**a**) and identified visco-elastic characteristics of the helical spring (**b**), hydraulic shock-absorber (**c**) and friction of the suspension mechanism (**d**)

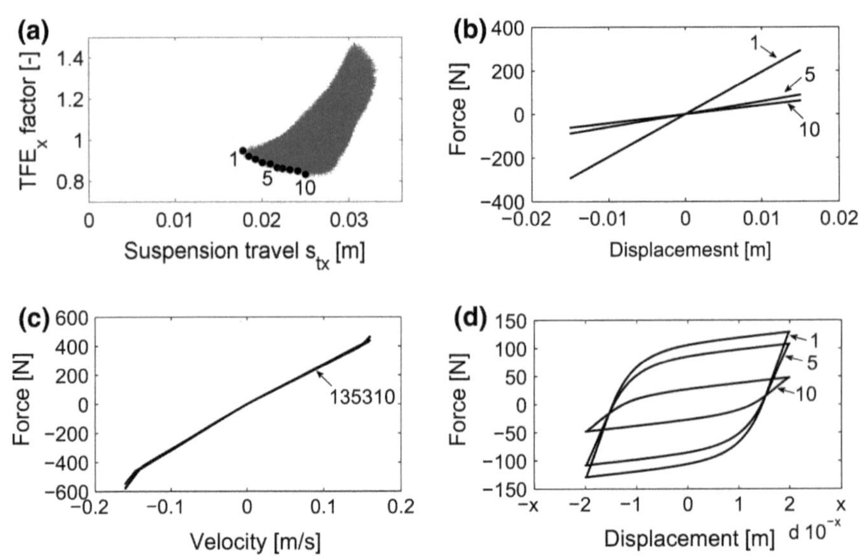

Fig. 5.9 Pareto-optimal point distribution for the input vibration WN3x (**a**) and identified visco-elastic characteristics of the helical spring (**b**), hydraulic shock-absorber (**c**) and friction of the suspension mechanism (**d**)

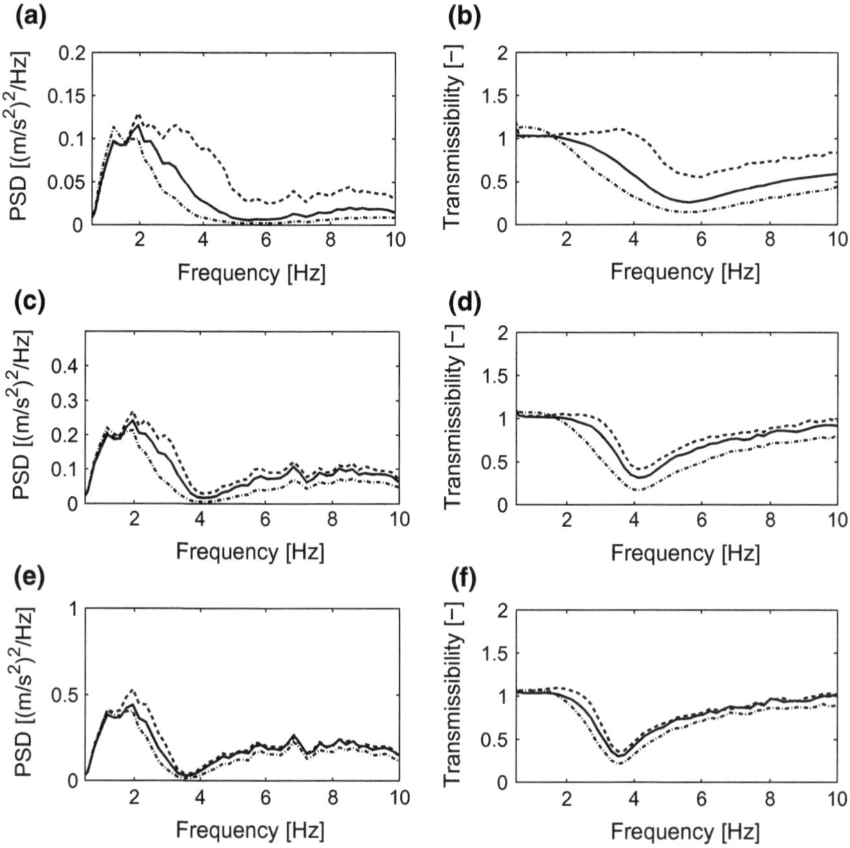

Fig. 5.10 Power spectral densities of the Pareto-optimal horizontal seat suspension at different system configurations: point No. 1 (----) , point No. 5 (——) , point No. 10 (-··-·) for various excitation intensities: WN1x (**a**), WN2x (**c**), WN3x (**e**) and transmissibility functions of the Pareto-optimal horizontal seat suspension at different system configurations: point No. 1 (----) , point No. 5 (——) , point No. 10 (-··-·) for various excitation intensities: WN1x (**b**), WN2x (**d**), WN3x (**f**)

characteristics of the helical springs, hydraulic shock-absorber and friction of the suspension mechanism. The particular system configurations regarding the marginal and compromising solutions (point No. 1, 5 and 10) are presented in Figs. 5.7b, c, d, 5.8b, c, d and 5.9b, c, d. These desired characteristics are defined by the values of the chosen decision variables: d_s, d_o, l_o and δ_f. Power spectral densities and transmissibility functions of the optimised seat suspension at various possibilities for the system configuration are shown in Fig. 5.10.

5.8 Experimental Research of the Optimised Horizontal Suspension with Seated Human Body

The experimental set-up used for evaluating system characteristics of the horizontal seat suspension is shown in Fig. 5.11a. During laboratory research, the conventional and optimal systems are tested in terms of vibration damping effectiveness. According to the optimisation results presented in previous section (Sect. 5.7), lower stiffness, harder damping and reduced friction are applied in the optimised horizontal suspension. Such a system configuration corresponds to the Pareto-optimal solution No. 10 that is obtained for the excitation signals at different intensities (Figs. 5.7, 5.8 and 5.9). For this solution, the smallest vibration amplitudes are measured on the seat; therefore, the transmissibility factor TFE_x shows the lowest values. The values of suspension travel s_{tx} do not exceed unacceptable stroke of the system; thus, the contact with end-stop buffers may be occurred occasionally. Laboratory evaluation of the operator seat vibration in longitudinal direction is performed using the experimental set-up presented in Fig. 5.11b.

The stiffness characteristics of the helical springs are evaluated by using quasi-static measurement of the force and the relative displacement of suspension system (Fig. 5.12a). During the measurement, a slow movement of the horizontal seat suspension is generated with the triangle cycling at a frequency of 0.1Hz and an amplitude of ±0.015m. Two different shock-absorbers are measured in order to estimate their damping forces. In this case, sinusoidal cycling of the dampers at various frequencies, i.e. 0.83 and 1.66 Hz, and an amplitude of ±0.0125 m are applied (Fig. 5.12b). The friction force of conventional and optimal suspension mechanisms is evaluated again with triangle cycling, however, at a frequency of 1 Hz and an amplitude of ±0.002 m (Fig. 5.12c). Such measurement conditions provide oscillating motion of the suspension mechanism, and the kinetic friction can be determined successfully.

Fig. 5.11 Experimental set-up for the force–deflection measurement of horizontal seat suspension (**a**) and test person with the seated human body (**b**)

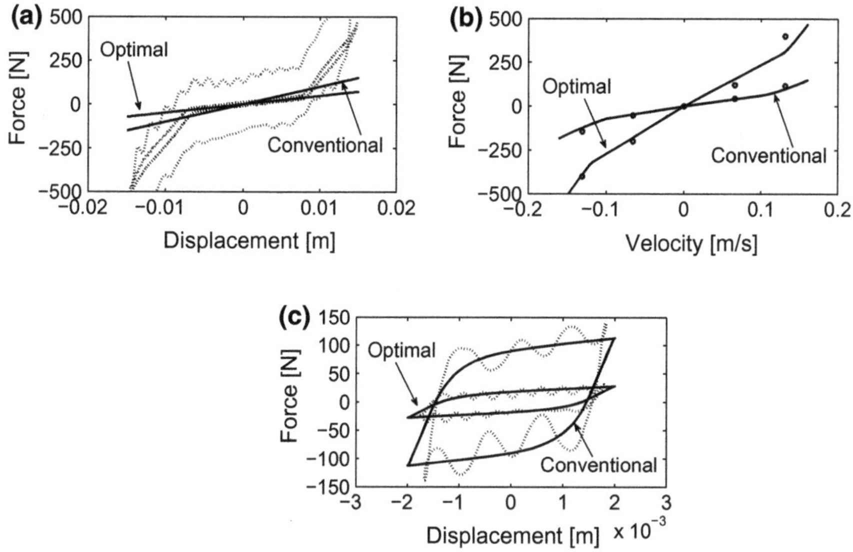

Fig. 5.12 Forces of the helical spring (**a**), hydraulic shock-absorber (**b**) and kinetic friction of the suspension mechanism (**c**) evaluated by means of simulation model (—), force–deflection measurement (''''''''') and force–velocity measurement of the damper (o)

Power spectral densities and transmissibility functions regarding serially manufactured and optimised seat suspensions are shown in Fig. 5.13. This figure clearly shows that the optimal horizontal suspension system significantly decreases the harmful influence of vibration acting on the human subject at various intensities of the chosen excitation signal. Much lower vibration amplitudes are discovered for the optimised seat suspension compared with the conventional system; and moreover, such a satisfying system behaviour is recognised in whole frequency range, i.e. starting from 0.5 Hz up to the frequency of 10 Hz.

Particular values of the transmissibility factor and suspension travel achieved for the following excitation signals: WN1x, WN2x and WN3x are listed in Table 5.1. The obtained vibro-isolation criteria definitely indicate the enhanced suspension performance as a result of the optimal system characteristics. The transmissibility factors TFE_x and the suspension travels s_{tx} are decreased simultaneously of about 15% in comparison with conventional system. It confirms that the proposed multi-criteria optimisation is properly developed for the purpose of improving the vibro-isolation properties of suspension systems. In order to better present the obtained results, the particular vibro-isolation criteria are presented as a bar graph in Fig. 5.14.

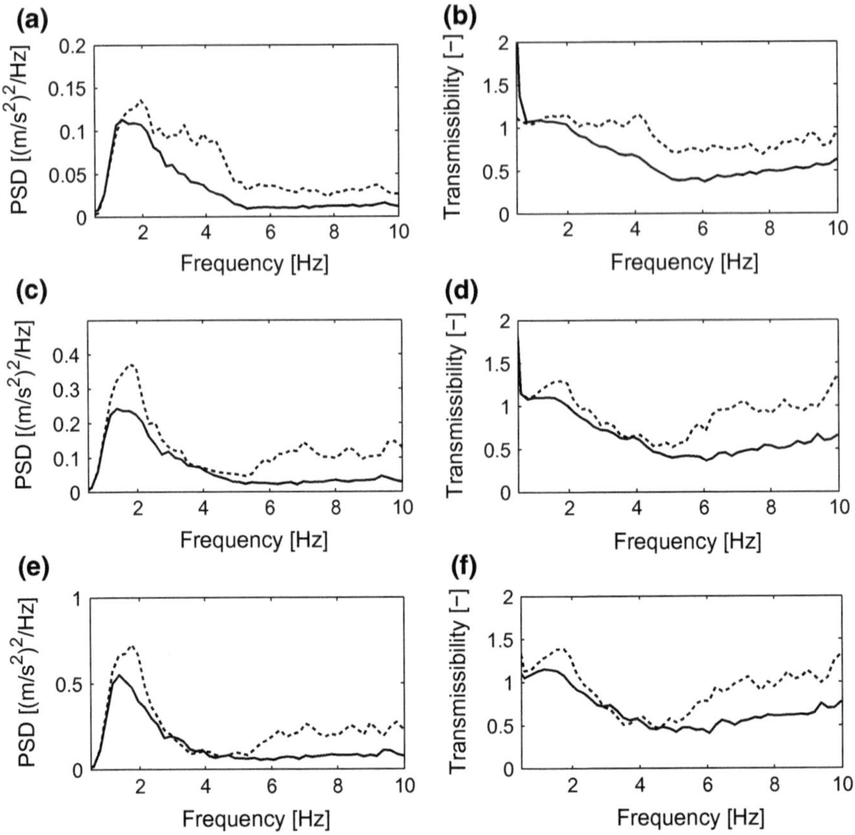

Fig. 5.13 Power spectral densities of the conventional (----) and optimal (—) horizontal seat suspension for different spectral classes: WN1x (**a**), WN2x (**c**), WN3x (**e**) and transmissibility functions of the conventional (----) and optimal (—) horizontal seat suspension for different spectral classes: WN1x (**b**), WN2x (**d**), WN3x (**f**), human body mass 90 kg

Table 5.1 Numerical values of the transmissibility factors and suspension travels obtained for excitation signals: WN1x, WN2x and WN3x, human body mass 90 kg

	Conventional		Optimal	
Input vibration	TFE_x factor	$s_{tx} mm,$	TFE_x factor	s_{tx}, mm
WN1x	1.050	18.2	0.900	14.8
WN2x	1.046	25.1	0.884	21.7
WN3x	1.071	28.2	0.899	25.9

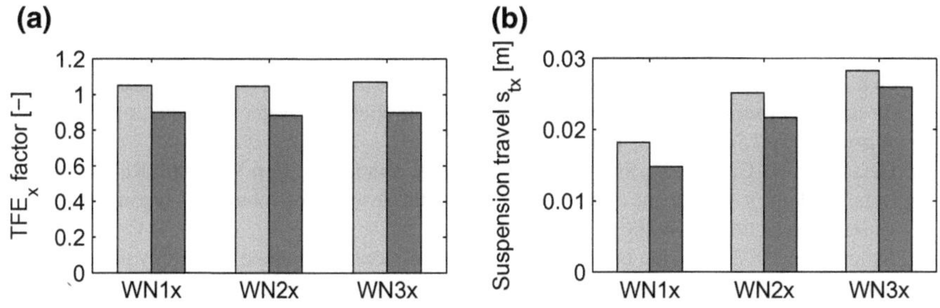

Fig. 5.14 Measured transmissibility factors and suspension travels of the conventional (■) and optimal (■) horizontal seat suspension with seated human body obtained for excitation signals: WN1x, WN2x and WN3x, human body mass 90 kg

5.9 Conclusions

In the following chapter an efficient methodology of selecting the vibro-isolation properties of vibration reduction systems is discussed. The global sensitivity analysis is applied in order to establish the global sensitivity indices of system performance due to many model parameters. Then, the multi-criteria optimisation is developed to determine Pareto-optimal system characteristics with respect to the conflicted vibro-isolation criteria, i.e. the frequency-weighted transmissibility factor and suspension travel. The usefulness of the proposed methodology is verified by performing experimental research of an exemplary horizontal seat suspension. Its visco-elastic characteristics are optimised uniquely for three different excitation intensities. In this case, acceptable compatibility of the experimental and simulation results is achieved for the investigated vibration reduction system.

References

1. Preumont, A. (2002). *Vibration control of active structures an introduction*. London: Kluwer Academic Publishers.
2. Kowal, J. (1996). *Vibration control (in Polish)*. Krakow: Gutenberg.
3. Censor, Y. (1977). Pareto optimality in multiobjective problems. *Applied Mathematics Optimization*, *4*, 41–59.
4. Marler, R. T., & Arora, J. S. (2004). Survey of multi-objective optimization methods for engineering. *Structural and Multidisciplinary Optimization*, *26*, 369–395.
5. Maciejewski, I. (2012). *Shaping the vibro-isolation properties of vibration reduction systems used for protection of working machines operators (in Polish)*. Koszalin: Academic Publishing of the Koszalin University of Technology.
6. Tarnowski, W. (2009). *Optimisation and polyoptimisation in mechatronics (in Polish)*. Koszalin: Academic Publishing of the Koszalin University of Technology.
7. Tarnowski W., Krzyzynski T., Maciejewski I., Olekiewicz R. (2011). *Poly-optimization: A paradigm in engineering design in mechatronics* (pp. 141–156), Archive of Applied Mechanics 81, Springer.

8. Pianosi, F., Sarrazin, F., & Wagener, T. (2015). A matlab toolbox for global sensitivity analysis. *Environmental Modelling and Software*, *70*, 80–85.
9. Goh, E. G., & Noborio, K. (2014). Sensitivity analysis using sobol 'variance-based method on the haverkamp constitutive functions implemented in Richards' water flow equation. *Malaysian Journal of Soil Science*, *18*, 19–33.
10. Dejaegher, B., Capron, X., Smeyers-Verbeke, J., & Vander, Heyden Y. (2006). Randomization tests to identify significant effects in experimental designs for robustness testing. *Analytica Chimica Acta*, *564*, 184–200.
11. Saltelli, A. (2002). Making best use of model evaluations to compute sensitivity indices. *Computer Physics Communications*, *145*, 280–297.
12. Saltelli, A., & Annoni, P. (2010). How to avoid a perfunctory sensitivity analysis. *Environmental Modeling and Software*, *25*, 1508–1517.
13. Saltelli, A., Annoni, P., Azzini, I., Campolongo, F., Ratto, M., & Tarantola, S. (2010). Variance based sensitivity analysis of model output. *Design and Estimator for the Total Sensitivity Index, Computer Physics Communications*, *181*, 259–270.
14. Maciejewski, I. (2012). Control system design of active seat suspensions. *Journal of Sound and Vibration*, *331*, 1291–1309.

Chapter 6
Computational Method of Selecting Vibro-isolation Properties

6.1 Nomenclature

$\mathbf{C}_i$	stiffness matrix of the bio-mechanical model of human body (i $= x, y, z$)
$\mathbf{D}_i$	damping matrix of the bio-mechanical model of human body (i $= x, y, z$)
F_{aij}	applied force of the active element (i $= x, y, z$), N
F_{cij}	applied force of the conservative element (i $= x, y, z$), N
F_{dij}	applied force of the dissipative element (i $= x, y, z$), N
$G_{HPij}(s)$	transfer function of the Butterworth high-pass filter
$G_{LPij}(s)$	transfer function of the Butterworth low-pass filter
$H_{hi}(s)$, $H_{li}(s)$	transfer functions of the filters limiting the scope of the analysed vibration frequency
$H_{ti}(s)$, $H_{si}(s)$	transfer functions of the filters shaping the characteristics of frequency weightings
$\mathbf{M}_i$	inertia matrix of the bio-mechanical model of human body (i $= x, y, z$)
PSD_{si}	power spectral density of the input vibration for selected direction of the vibration exposure (i $= x, y, z$), $(\mathrm{m/s^2})^2/\mathrm{Hz}$
S_j	first-order sensitivity index
S_{Tj}	total effect sensitivity index
T_i	transmissibility function for selected direction of the vibration exposure (i $= x, y, z$)
TFE_i	frequency-weighted transmissibility factor for selected direction of the vibration exposure (i $= x, y, z$)
VDV_i	vibration dose value, $\mathrm{m/s^{1.75}}$
a_{RMS}	root mean square value of the acceleration signal, $\mathrm{m/s^2}$
f	frequency, Hz
f_{HPij}	cut-off frequency of the high-pass filter, Hz
f_{LPij}	cut-off frequency of the low-pass filter, Hz

T. Krzyzynski et al., *Modelling and Control Design of Vibration Reduction Systems*, https://doi.org/10.1007/978-3-030-03047-6_6

k_{i}	multiplying factor for selected direction of the vibration exposure (i $= x, y, z$)
$q_{1\mathrm{i}}$	displacement of the isolated body for selected direction of the vibration exposure (i $= x, y, z$), m
q_{si}	displacement of the input vibration (i $= x, y, z$), m
s	Laplace variable
s_{ti}	suspension travel for selected direction of the vibration exposure (i $= x, y, z$), m
t	time, s
$\mathbf{x}_{\mathrm{d}}$	vector of the selected decision variables
σ_{ij}^2	variance of the random numbers

6.2 Introduction

The dynamical properties of vibration reduction systems are usually estimated by using the integral criteria that are measured during the vibro-isolation process. For this reason, an appropriate selection of the system characteristics is realised by using a computer model to simulate the system dynamics under well-defined operating conditions. A method of selecting vibro-isolation properties would support to discover the essential visco-elastic characteristics that successively can be modified in order to achieve the best system performance in terms of its conflicted requirements. The study presented in this chapter is conducted for the needs of obtaining the high reduction of mechanical vibrations transmitted to the human body while simultaneous limiting of the suspension travel.

6.3 Formulation of the Overall Method

The block diagram of the overall method for selecting vibro-isolation properties of vibration reduction systems is presented in Fig. 6.1. Assuming that the input vibration is provided by means of a random signal generator together with a suitable filtration technique, the vibro-isolation properties of the system under study would be determined by using its simulation model. The visco-elastic characteristics of vibration reduction system must be evaluated in relation to the opposite system requirements. Therefore, the multi-criteria optimisation procedure is required in order to select the non-linear system configuration which satisfies the conflicted vibro-isolation criteria.

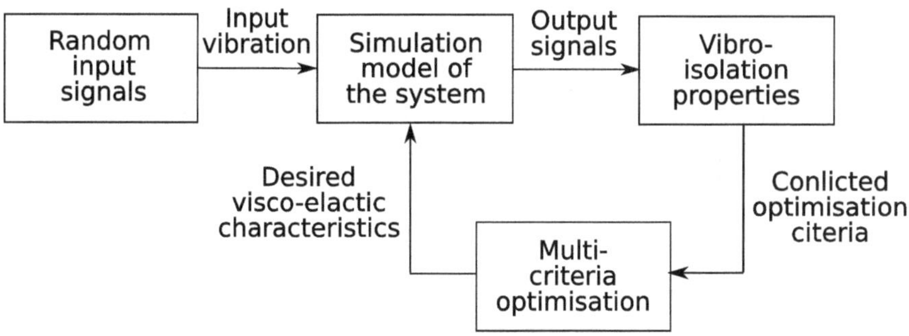

Fig. 6.1 Block diagram of the overall method for selecting vibro-isolation properties of vibration reduction systems

6.4 Random Input Signals

In order to analyse the dynamics of vibration isolation systems, the excitation signal with specific spectral characteristics should be used. Unfortunately, there are no effective methods for generating the signals representing the work of different machineries. In the papers [1–3], the target spectral characteristics are only standardised for the simulated input vibration test in vertical direction (Fig. 6.2a). There is a lack of effective procedures for generating the time history of random signals with precisely defined spectral properties (Fig. 6.2b). In addition, it is not known how to reproduce the excitation signals based on the measured spectral characteristics [4], especially in horizontal directions (lateral and longitudinal).

In the following section, an efficient method of generating the random vibration with strictly defined spectral classes is proposed in order to select the dynamic characteristics of vibration reduction systems. The elaborated models of input signals effectively support a testing of the vibration reduction systems by using the laboratory measurements and/or computer simulations. The proposed method can be

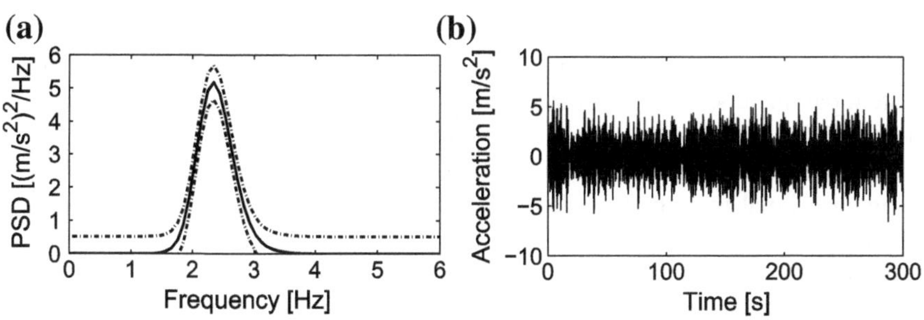

Fig. 6.2 Target spectral characteristics of the input vibration (**a**) and the time history of random signal with precisely defined spectral properties (**b**)

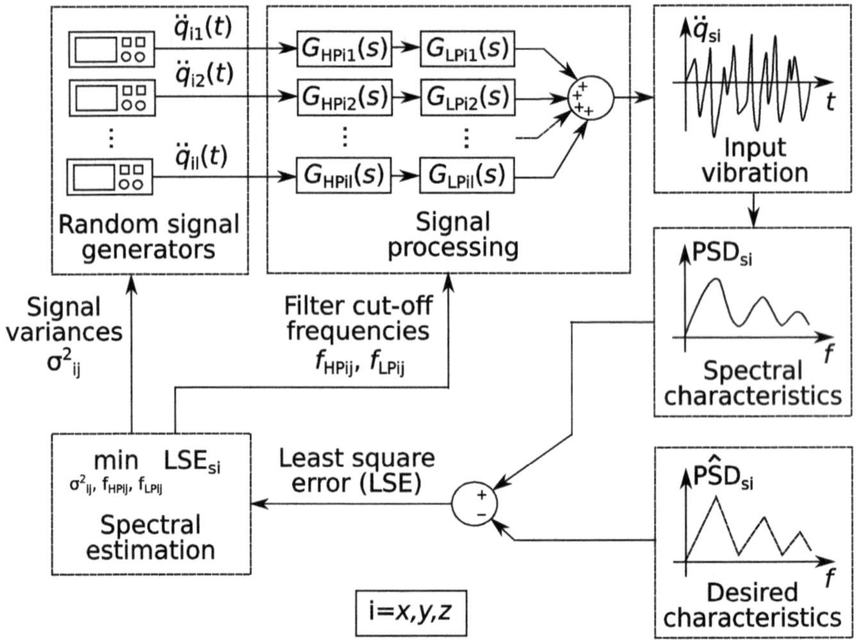

Fig. 6.3 Block diagram of modelling the vibration exposure in typical working machines by means of random input signals

used for generating the signals specified by present test standards [1–3] and also for reproducing the signals measured by other authors [4]. The block diagram of modelling the vibration exposure in typical working machines by means of random input signals is presented in Fig. 6.3.

As shown in Fig. 6.3, at first the set of normally distributed random signals $\ddot{q}_{1i}(t)$, $\ddot{q}_{2i}(t)$,...,$\ddot{q}_{li}(t)$ should be generated (see Sect. 2.3) for a specific direction of vibration exposure (i $= x, y, z$). Then, the high-pass $G_{HPi1}(s)$, $G_{HPi2}(s)$,..., $G_{HPij}(s)$ and low-pass $G_{LPi1}(s)$, $G_{LPi2}(s)$,..., $G_{LPij}(s)$ filters of the Butterworth type (Eqs. 2.3–2.4) are used to achieve the input vibration $\ddot{q}_{si}$ with a precisely defined power spectral density PSD_{si}. The spectral estimation method is proposed in order to evaluate the unknown system parameters, i.e. signal variances σ_{ij}^2 and filter cut-off frequencies f_{HPij}, f_{LPij} (Eqs. 2.9–2.11). Such a parametric method allows to find the system configuration with magnitude response approximating a desired spectral characteristics. Therefore, a random signal with the user-defined spectral properties can easily be generated. The proposed method applies the least square error (LSE) minimisation technique over the frequency range of the filter's response (see Sect. 2.5).

6.5 Simulation Model of the Passive System

For the purpose of computing the dynamic response of a vibration reduction system to specialised input vibrations, the simulation model should be utilised. Hence, the ordinary differential equations describing system dynamics shall be solved numerically with the high accuracy and precision [5]. Such a procedure provides fast prototyping of the system configuration, while various parameters of the model are investigated iteratively. The block diagram representing a general simulation model of the passive vibration reduction system is presented in Fig. 6.4.

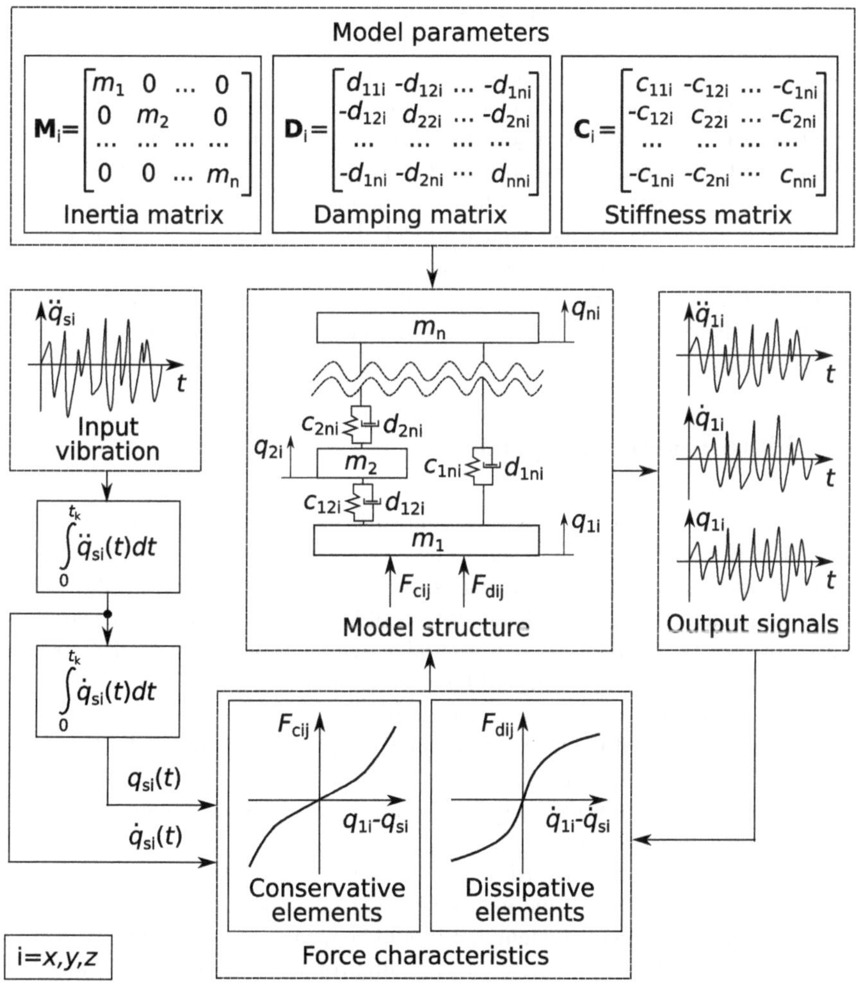

Fig. 6.4 Block diagram for numerical simulation of the dynamic behaviour of passive vibration reduction systems

The input vibration block (Fig. 6.4) is used in order to bring about the acceleration signal $\ddot{q}_{si}(t)$ for a single axis of the vibration transmission (i $= x, y, z$). For the reason that the simulation model demands the velocity $\dot{q}_{si}(t)$ and displacement $q_{si}(t)$ as input signals, the acceleration signal has to be integrated over the time t prior to adopting it as the kinematic excitation. In turn, the conservative F_{cij} and dissipative F_{dij} forces (Eq. 3.5) of principal elements included in the vibration reduction system must be modelled as functions of the relative velocity $\dot{q}_{1i} - \dot{q}_{si}$ and relative displacement $q_{1i} - q_{si}$. Their non-linear characteristics should be determined, especially for the components utilised in a specific suspension type (see Sect. 3.4).

Further modelling procedure concerns selecting a model structure which is reliable for representing the human behaviour exposed to whole-body vibration. There is a great amount of the human body models available in the current literature [6–8]; nonetheless, exemplary models of the seated body under vertical and horizontal vibration are discussed in Sect. 3.7. The set of independent equations of motion results from the chosen model structure and shall be depicted in the matrix form (Eq. 3.1). The model parameters are used to determine particular elements of the inertia $\mathbf{M}_i$, damping $\mathbf{D}_i$ and stiffness $\mathbf{C}_i$ matrices (Eqs. 3.7–3.11). Their values influence the system dynamics and therefore define individual body displacements q_{1i}, q_{2i}, …, q_{ni} of the bio-mechanical model. The output signal block is recommended among others in order to measure isolated body acceleration $\ddot{q}_{1i}(t)$. The other signals, i.e. the velocity $\dot{q}_{1i}(t)$ and displacement $q_{1i}(t)$, are utilised for shaping the characteristics of visco-elastic elements contained in a suspension system.

6.6 Vibro-isolation Properties

Time history of the input acceleration (Fig. 6.5a) and equivalent vibration of the isolated body (Fig. 6.5b) must be measured for the purpose of calculating the vibro-isolation criteria. Even if statistical analysis of the random waveforms leads to some signal parameters, the spectral characteristics of human vibration shall be estimated due to a different human sensitivity at a variable frequency of motion [9]. The spectral content is analysed by applying the fast Fourier transform (FFT) of time history signals [10]. The power spectral density (PSD) is utilised to present and compare the frequency spectra of input acceleration (Fig. 6.5c) and isolated body (Fig. 6.5d) exposed to random vibration [11].

For the first system evaluation, the dynamic properties of vibration reduction systems are analysed using the block diagram shown in Fig. 6.6. As shown in this figure, a three-channel data acquisition system is used to record the following signals: the input acceleration $\ddot{q}_{si}(t)$, the isolated body acceleration $\ddot{q}_{1i}(t)$ and the relative displacement of suspension system that should be measured for a specific direction of the vibration exposure (i $= x, y, z$). A qualitative assessment of the relationship between the input $\ddot{q}_{si}(t)$ and output $\ddot{q}_{1i}(t)$ acceleration signals is performed on the basis of power spectral densities in the frequency domain f. Then, the transmissibility

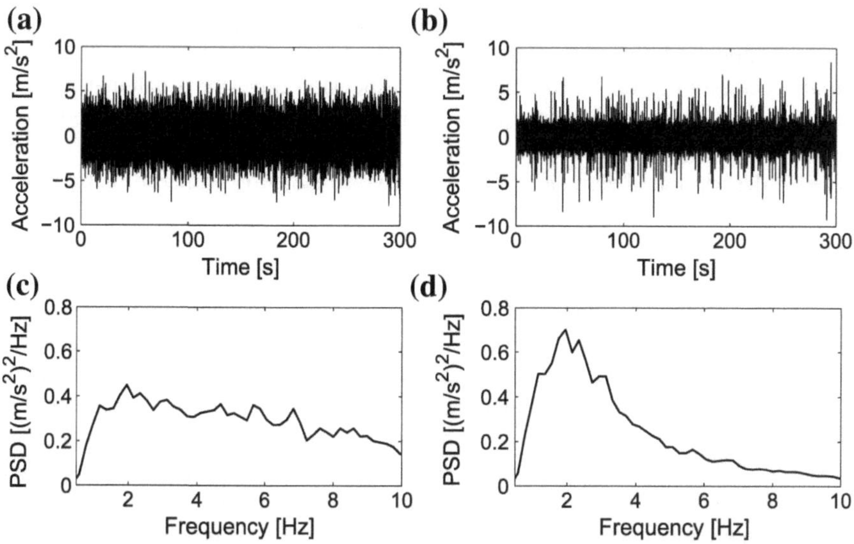

Fig. 6.5 Exemplary time history of the input acceleration (**a**) and corresponding vibration of the isolated body, power spectral densities of the input acceleration (**c**) and corresponding isolated body vibration (**d**)

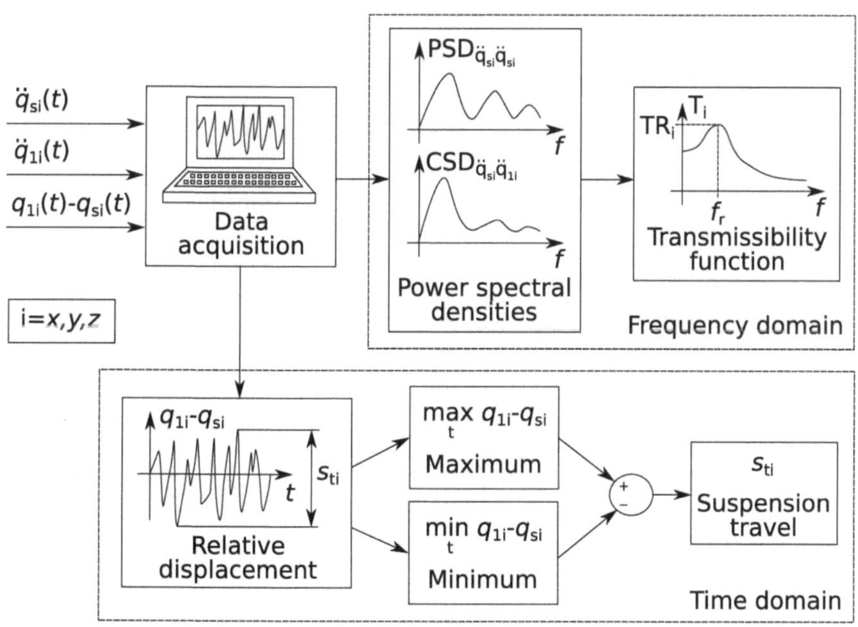

Fig. 6.6 Block diagram for evaluating the dynamic properties of vibration reduction systems

function T_i is estimated (Eq. 4.13) as a quotient of the cross-spectral density $CSD_{\ddot{q}_{si}\ddot{q}_{1i}}$ between input and output accelerations and the power spectral density $PSD_{\ddot{q}_{si}\ddot{q}_{si}}$ of input acceleration itself. Furthermore, the suspension travel s_{ti} is calculated based on the time history of displacement signal $q_{1i}(t) - q_{si}(t)$ and the achieved value expresses the highest deflection (rebound) of suspension relative to its base. The detailed description of estimating the system vibro-isolation properties is presented in Sect. 4.4.

For the second system evaluation, the frequency-weighting method [12] is employed to determine the risks coming from human exposure to mechanical vibration. Such a method assumes that the risk arising from the whole-body vibration is significant in the frequency range 0.5–80 Hz, but the vibration risk is not equal for all frequencies contained in that range. As a consequence of the frequency weighting, an occurrence of the low frequencies is emphasised and the weighting factor decreases when the frequency increases [13]. However, the risks from vibration depend also on the duration of exposure. Therefore, the daily vibration exposure should be calculated based on both the level of vibration and the exposure time. The block diagram for assessing the human exposure to whole-body vibration is presented in Fig. 6.7.

There are four filter sections (Eqs. 4.1–4.4) required in order to perform the frequency weightings of the input vibration $\ddot{q}_{si}(t)$ and the isolated body acceleration $\ddot{q}_{1i}(t)$. The first two filter components ($H_{hi}(s)$ and $H_{li}(s)$) are used to restrict

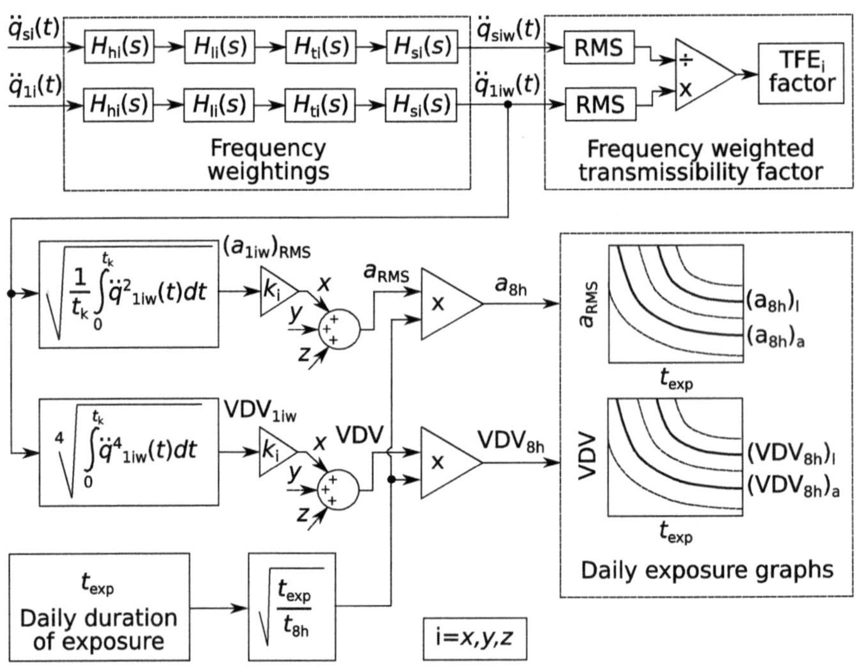

Fig. 6.7 Block diagram for assessing the human exposure to whole-body vibration

the frequency range that should be analysed for whole-body vibration influence on the human organism. The remaining components ($H_{ti}(s)$ and $H_{si}(s)$) are utilised for shaping the transfer characteristics of a total frequency-weighting filter that are applicable for a single axis of the vibration transmission ($i = x, y, z$) and various positions of the human body at work (sitting or standing). Principal frequency weightings for the exposure evaluation according to ISO 2631 are presented in Table 4.1. The transmissibility factor TFE_i is then calculated (Eq. 4.15) as a ratio between the frequency-weighted root mean square acceleration $\ddot{q}_{liw}(t)$ transmitted to the human body and the measured acceleration $\ddot{q}_{siw}(t)$ of input vibration.

For the further analysis, the duration of exposure to the whole-body vibration is taken into account because the longer duration of exposure typically causes the greater risk to health arising from vibrations. In this case, the root mean square acceleration $(a_{liw})_{RMS}$ and the vibration dose value VDV_{liw} must be calculated (Eqs. 4.7–4.8) for each individual axis ($i = x, y, z$). Then, the frequency-weighted vibration values have to be multiplied by the additional factors k_i [13]. For the horizontal axes (x and y), this factor is equal to 1.4, and for the vertical vibration (z-axis), the factor is 1.0. The vibration total value of weighted root mean square acceleration a_{RMS} (Eq. 4.9) and the total vibration dose value VDV (Eq. 4.10) that are determined from vibration in three orthogonal coordinates are related to the daily duration t_{exp} of exposure. Finally, the root mean square acceleration a_{8h} (Eq. 4.11) and the vibration dose value VDV_{8h} (Eq. 4.12) calculated during a working day should be compared with the daily exposure action ($(a_{RMS})_a$ and $(VDV_{8h})_a$) and limit values ($(a_{RMS})_l$ and $(VDV_{8h})_l$) by using the daily exposure graphs. The evaluation of the human exposure to whole-body vibration is exhaustively discussed in Sect. 4.3.

6.7 Multi-criteria Optimisation of the Vibro-isolation Properties

In order to optimise the vibro-isolation properties of vibration reduction systems, the global sensitivity analysis is employed at first. By using such analysis, the relative importance of input factors with respect to model output should be quantified to identify the key model parameters that affect the system performance significantly. The variance-based method [14] is used in the presented study because such a method is suitable for non-linear models that are generally used to describe a specific vibro-isolation process realised by the system [15]. Block diagram illustrating the sensitivity analysis of vibration reduction systems is shown in Fig. 6.8.

Based on both the selected number N of simulation runs and the double number 2p of input factors to be analysed, the matrix $\mathbf{X}_{AB}$ of initial starting points should be generated by using a uniform random number generator (see Sect. 5.4.1). Then, the resulting sample matrix has to be divided into two separated matrices $\mathbf{X}_A$ and $\mathbf{X}_B$ of the same size ($N \times$ p) that include two independent sets of initial starting points.

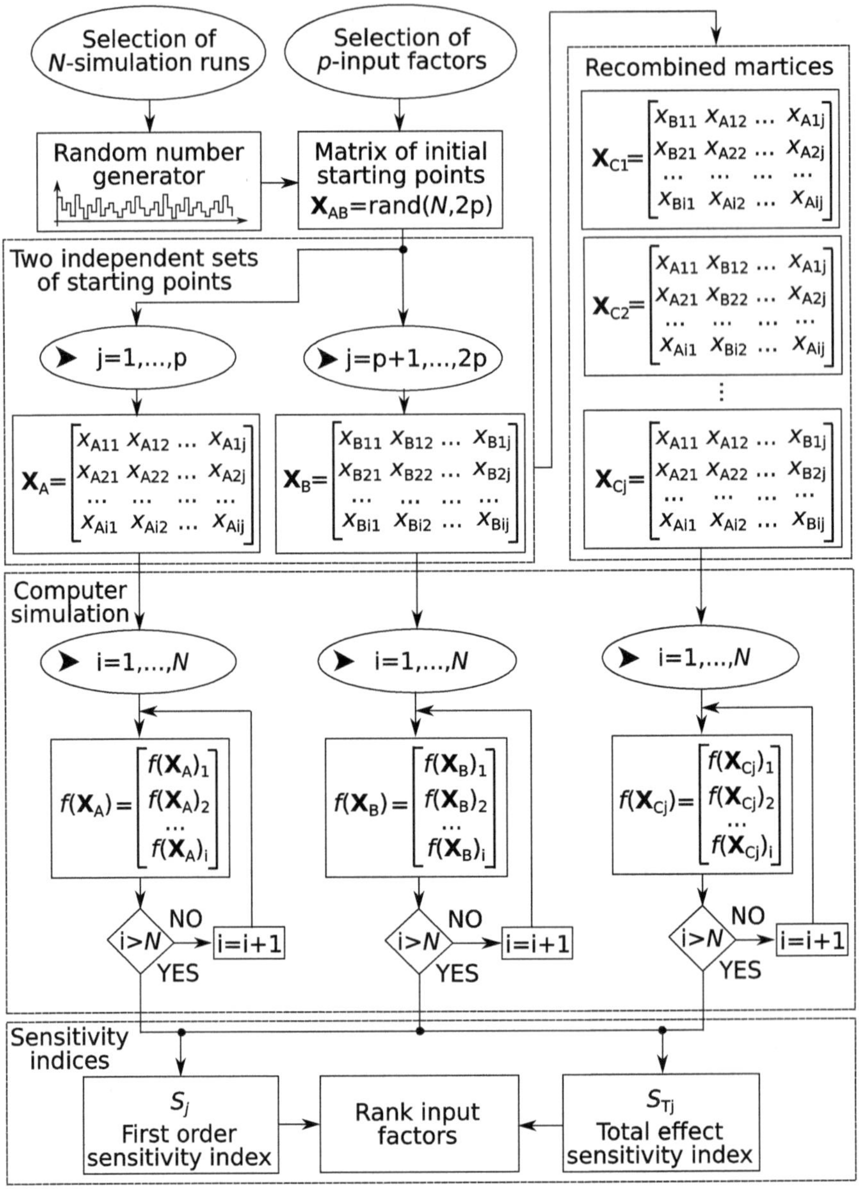

Fig. 6.8 Block diagram for the sensitivity analysis of vibration reduction systems

Additionally, the recombined matrices $\mathbf{X}_{C1}, \mathbf{X}_{C2}, \ldots, \mathbf{X}_{Cj}$ must be created in such a way that these matrices contain the elements of matrix $\mathbf{X}_A$ except the jth column which is taken from matrix $\mathbf{X}_B$. Further analysis requires $N(2+\text{p})$ model calculations by using a computer simulation. The corresponding model outputs have to be

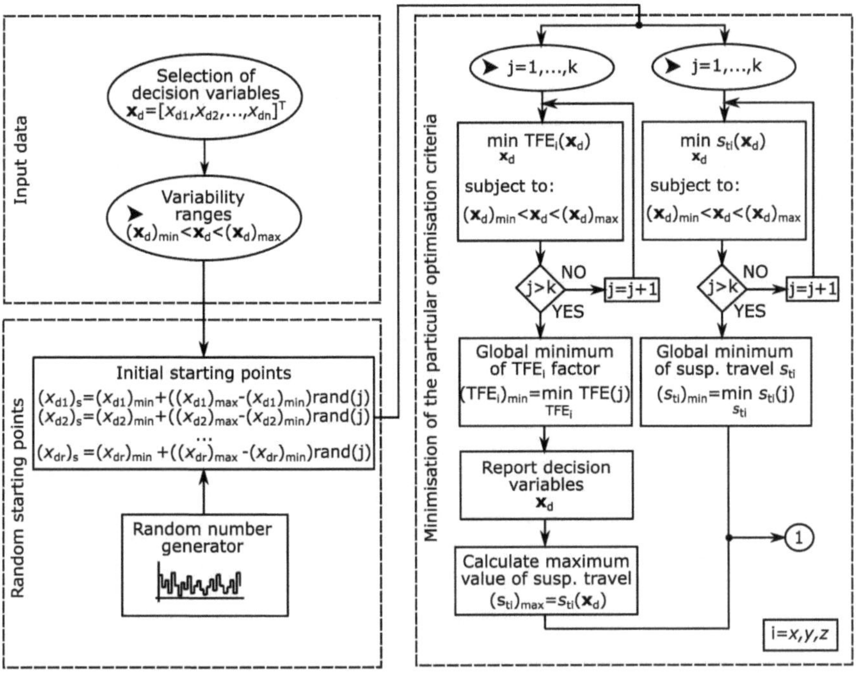

Fig. 6.9 Block diagram for the optimisation procedure of vibration reduction systems—part 1

stored as the following vectors: $f\ (\mathbf{X}_A)$, $f\ (\mathbf{X}_B)$ and $f\ (\mathbf{X}_{Cj})$. A detailed description of the variance-based method is presented in Sect. 5.4.2. When all of the model outputs are iteratively evaluated, then the sensitivity indices shall be estimated. The first-order sensitivity index S_j and the total effect sensitivity index S_{Tj} are calculated using Eqs. (5.12)–(5.14), and the obtained results can be used to provide a reliable ranking of the input factors.

If the input factors are ranked according to their influence on the model output, then the parameters with the largest sensitivity indices should be collected as the vector $\mathbf{x}_d = [x_{d1}, x_{d2}, \ldots, x_{dn}]^T$ of decision variables for optimisation (Figs. 6.9 and 6.10). Their variability ranges can be determined as the vectors $(\mathbf{x}_d)_{min}$ and $(\mathbf{x}_d)_{max}$ that contain the minimum $(x_{d1})_{min}, (x_{d2})_{min}, \ldots, (x_{dn})_{min}$ and maximum $(x_{d1})_{max}, (x_{d2})_{max}, \ldots, (x_{dn})_{max}$ values of each decision variables, respectively.

In order to increase the probability of finding the global minimum of objective function, the random staring points are generated by using a uniformly distributed random number generator (see Sect. 5.6.2). Hence, a separate minimising of the

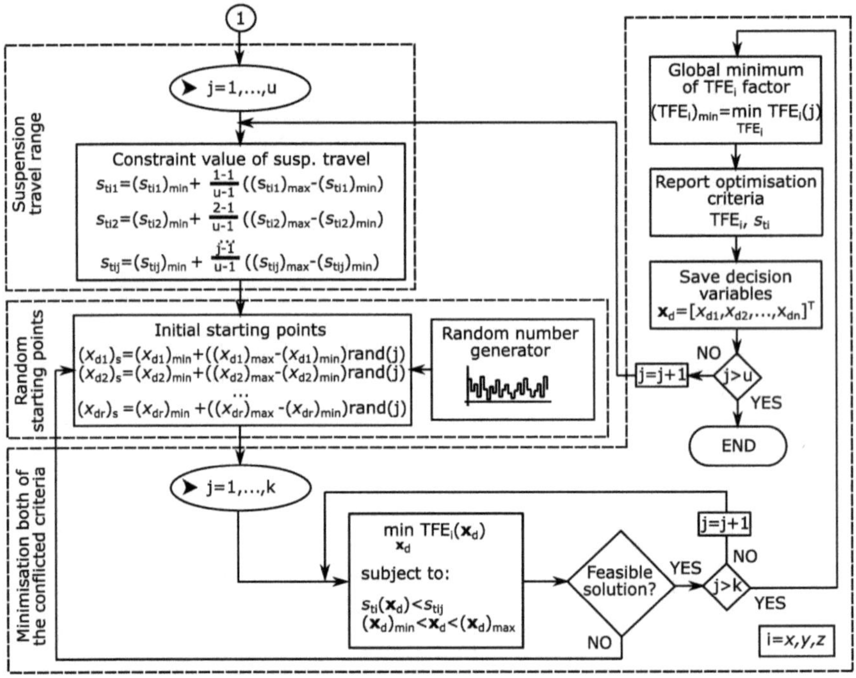

Fig. 6.10 Block diagram for the optimisation procedure of vibration reduction systems—part 2

particular optimisation criteria is realised repeatedly (Eqs. 5.19–5.21) until global minima of the frequency-weighted transmissibility factor $(\mathrm{TFE_i})_{\min}$ and suspension travel $(s_{\mathrm{ti}})_{\min}$ are found. In turn, the maximum value $(s_{\mathrm{ti}})_{\max}$ of suspension travel should be calculated in order to discover the range $(s_{\mathrm{ti}})_{\max} - (s_{\mathrm{ti}})_{\min}$ of suspension travel that belongs to the Pareto-front of a multi-criteria optimisation problem. Then, the linearly spaced vector $\mathbf{S}_{\mathrm{ti}} = [s_{\mathrm{ti1}}, s_{\mathrm{ti2}}, \ldots, s_{\mathrm{tij}}]$ of suspension travel should be defined for the purpose of the constraint method [16, 17]. Using random starting points, a minimising both of the conflicted optimisation criteria is carried out in such a way that the transmissibility factor $\mathrm{TFE_i}\,(\mathbf{x}_\mathrm{d})$ is chosen as an primary objective to be minimised. In turn, the suspension travel $s_{\mathrm{ti}}\,(\mathbf{x}_\mathrm{d})$ is treated as the non-linear inequality-constrained problem for the selected optimisation method (Eqs. 5.24–5.26). When the global minimum of the frequency-weighted transmissibility factor $(\mathrm{TFE_i})_{\min}$ is found for various constraints imposed on the suspension travel s_{tij}, then the design parameters are reported as corresponding values of the chosen decision variables and define visco-elastic characteristics of the suspension system.

6.8 Selection of the Visco-elastic Characteristics of Horizontal Seat Suspension for Chosen Excitation Signals

6.8.1 Random Vibration Inputs Affecting the Operators of Different Machineries at Work

The dynamical properties of horizontal seat suspension are adjusted to different spectral classes of the excitation signals that occur in the cabins of the following machinery: AL1x—articulated truck over rough ground, AT1x—agricultural tractor on road at speed 30 kph, AT2x—agricultural tractor on road at speed 40 kph, AT3x—agricultural tractor at speed 3 kph, AT4x—agricultural tractor at speed 20 kph and RV1x—rail vehicle. Using a general method to generate vibration signals of different machines (see Sect. 6.4), the created random signals are used as excitation signals for a vibration shaker as well as vibration inputs for a simulation model. Simulated and measured power spectral densities of the selected input vibrations are shown in Fig. 6.11.

The excitation signal with specific spectral characteristics (Fig. 6.11) is generated, especially for the longitudinal x-axis of the Cartesian coordinate system that corresponds to the direction of machine movement. There are some signals representing the work of different machineries with relatively low amplitudes of vibration, i.e. AT1x and RV1x, while the lowest excitation frequencies can be observed in the signal AL1x. The signals AL1x and AT2x have the medium amplitude range in the frequency range 0.5–4; however, the higher vibration amplitudes are obtained for the spectral classes AT3x and AT4x.

As shown in Fig. 6.11, spectral characteristics of the simulated input signals are very close to the characteristics measured experimentally. The lowest agreement between vibration amplitudes is shown for the low-intensity vibrations at the higher frequencies (more than 10 Hz). The very low vibration amplitudes measured in this frequency range are generated due to measurement noise affecting the acceleration signal. However, such low accelerations are insignificant in terms of determining the vibro-isolation properties of seat suspension systems.

6.8.2 Sensitivity Analysis of the Passive System for Various Visco-elastic Characteristics

The sensitivity analysis is performed by using a simulation model of the passive system (see Sect. 6.5). An influence of non-linear force characteristics of the conservative F_{cij} and dissipative F_{dij} elements is investigated for chosen excitation signals. Components applied in the horizontal seat suspension generate the following reaction forces: the helical spring force F_{cx1} (conservative), the force from buffers F_{cx2}

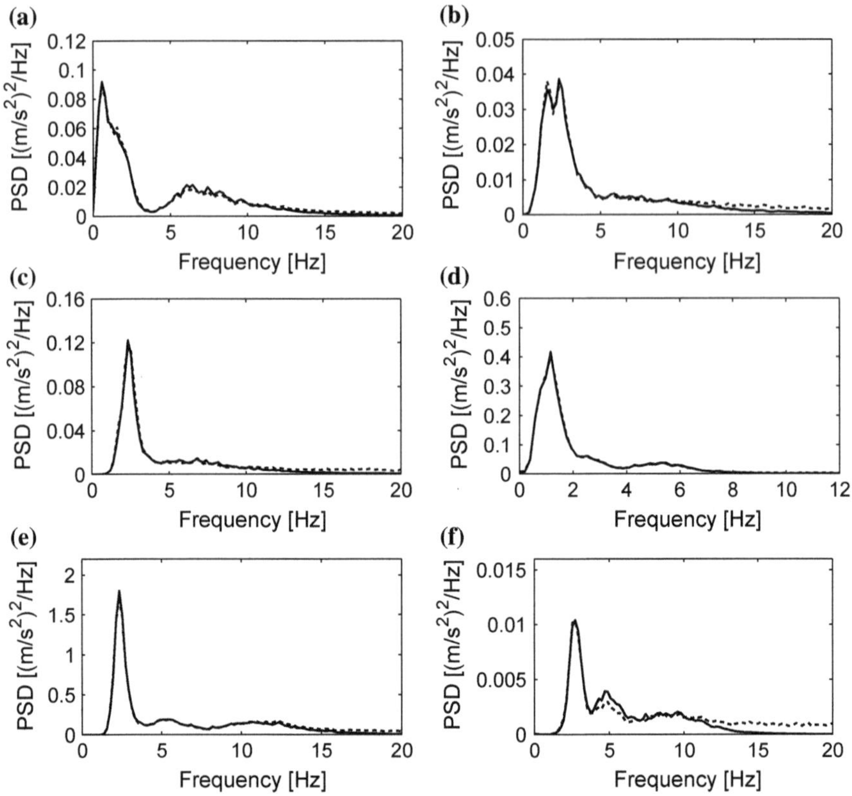

Fig. 6.11 Simulated (——) and measured (----) power spectral densities of chosen excitation signals: AL1x (**a**), AT1x (**b**), AT2x (**c**), AT3x (**d**), AT4x (**e**), RV1x (**f**)

(conservative), the force of hydraulic damper F_{dx1} (dissipative) and the friction force of suspension mechanism F_{dx2} (dissipative). Individual characteristics of the visco-elastic elements included in horizontal seat suspension are presented in Fig. 6.12.

The helical springs can be changed by modifying the wire diameter d_s; thus, its stiffness characteristics may be varied (Fig. 6.12a). The stiffness characteristics of end-stop buffers can be also adjusted using the linear c_{b1} (Fig. 6.12b) and cubic c_{b3} (Fig. 6.12c) stiffness coefficients. The damping characteristics of hydraulic shock-absorber may be managed by a different geometry of the orifice. The orifice diameter d_o influences the amount of damping force (Fig. 6.12d), while the orifice length l_o affects the non-linear damping characteristics (Fig. 6.12e). Finally, the friction of suspension mechanism may be customised by applying different values of the reduction ratio δ_f (Fig. 6.12f).

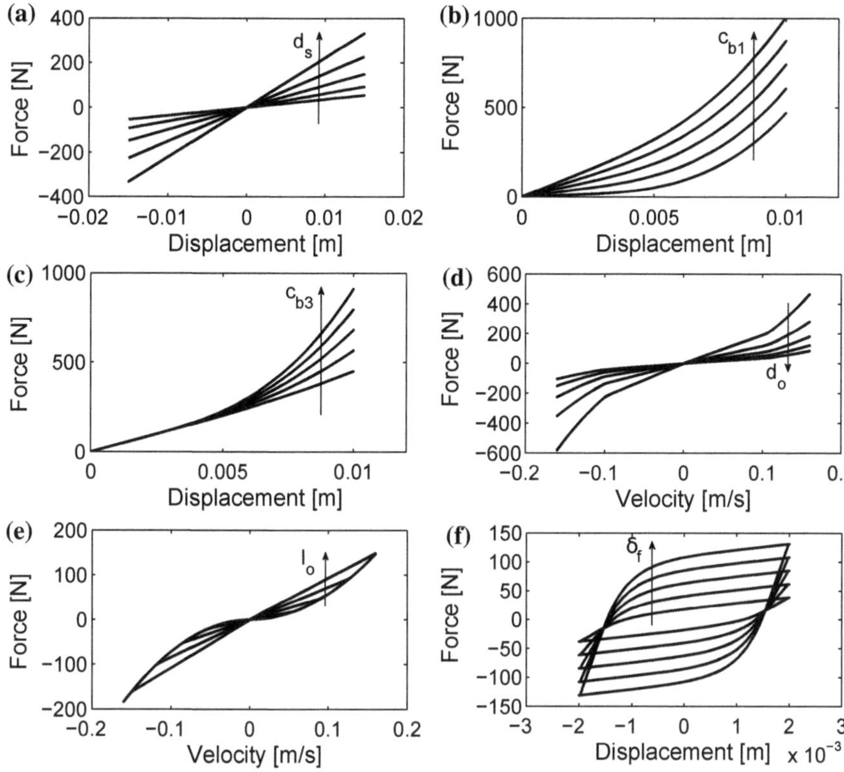

Fig. 6.12 Force characteristics F_{cx1} of the helical spring for various wire diameters d_s (**a**), force characteristics F_{cx2} of the end-stop buffer for various linear stiffness coefficients c_{b1} (**b**), force characteristics F_{cx1} of the end-stop buffers for various cubic stiffness coefficients c_{b3} (**c**), force characteristics F_{dx1} of the shock-absorber for various orifice diameters d_o (**d**), force characteristics F_{dx1} of the shock-absorber for various orifice lengths l_o (**e**), friction force F_{dx2} for various reduction ratios δ_f (**f**)

The vibro-isolation criteria, i.e. the frequency-weighted transmissibility factor $\mathrm{TFE_x}$ and the suspension travel s_{tx}, are evaluated on the basis of simulation results. The transmissibility factor is calculated using the following expression:

$$\mathrm{TFE_x} = \frac{(a_{1xw})_{RMS}}{(a_{sxw})_{RMS}} \tag{6.1}$$

where $(a_{1xw})_{RMS}$ is the frequency-weighted RMS acceleration of the suspended body measured for longitudinal x-axis of the vibration transmission and $(a_{sxw})_{RMS}$ is the frequency-weighted RMS acceleration of the input vibration. The frequency weightings defined in ISO 2631 [12] have to be used for the purpose of calculating the frequency-weighted accelerations (see Sect. 6.6).

The suspension travel is defined as the measure of distance from the bottom of the suspension stroke to the top of the suspension stroke using the following relation:

$$s_{tx} = \max_{t\in[0,t_k]} (q_{1x}(t) - q_{sx}(t)) - \min_{t\in[0,t_k]} (q_{1x}(t) - q_{sx}(t)) \tag{6.2}$$

where $q_{1x}(t)$ is the displacement of the suspended body measured for longitudinal x-axis of the vibration transmission, $q_{sx}(t)$ is the displacement of the vibrating platform, t is the present moment of time and t_k is the measurement time.

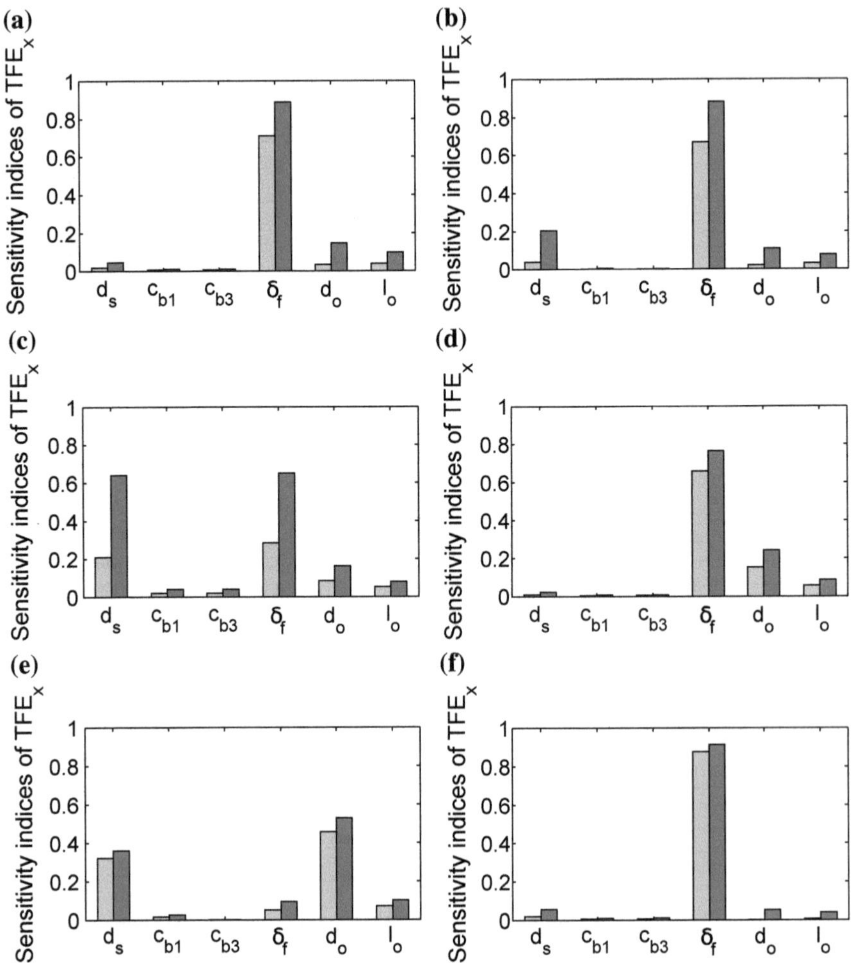

Fig. 6.13 First-order (light grey) and total effect (dark grey) sensitivity indices of transmissibility factor TFE_x for different input vibrations: AL1x (a), AT1x (**b**), AT2x (**c**), AL3x (**d**), AT4x (**e**), RV1x (**f**)

The first-order and total effect sensitivity indices of transmissibility factor TFE_x and suspension travel s_{tx} are calculated for chosen input vibrations, i.e. AL1x, AT1x, AT2x, AL3x, AT4x and RV1x (see Sect. 6.7). According to the results obtained by using the variance-based method (Figs. 6.13 and 6.14), the friction force (input factor δ_f) demonstrates the greatest influence on the system dynamics for most of the excitation signals. Less influence of the suspension friction is observed for the signal AT4 with relatively low frequencies and high amplitudes of vibration. In the case of such a signal, the damping force (input factor d_o) is dominant; therefore, the sensitivity indices achieve substantial values. Stiffness characteristics of the end-stop buffers (input factors c_{b1} and c_{b3}) point out the lowest influence on the system vibro-isolation properties for each excitations. For this reason, shaping the force characteristics F_{cx2} is omitted in the further investigations.

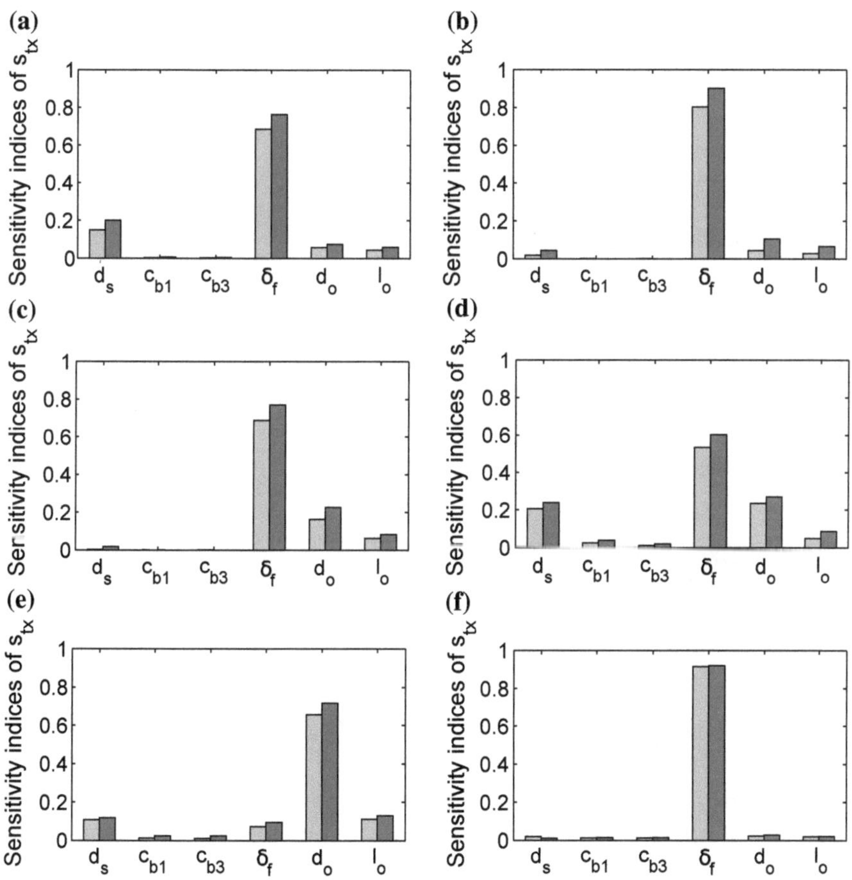

Fig. 6.14 First-order ( ) and total effect ( ) sensitivity indices of suspension travel s_{tx} for different input vibrations: AL1x (**a**), AT1x (**b**), AT2x (**c**), AL3x (**d**), AT4x (**e**), RV1x (**f**)

6.8.3 Multi-criteria Optimisation of the Vibro-isolation Properties

In reference to the results obtained in the previous section, there are four decision variables: the wire diameter $x_{d1} := d_s$ of helical spring, the orifice diameter $x_{d2} := d_o$ and length $x_{d3} := l_o$ of hydraulic damper and the reduction ratio $x_{d4} := \delta_f$ of friction force as important design parameters. Hence, the decision variables are settled in the following vector:

$$\mathbf{x}_d = [x_{d1}, x_{d2}, x_{d3}, x_{d4}]^T \tag{6.3}$$

The Pareto-optimal solutions of horizontal seat suspension are searched by iteratively minimising the frequency-weighted transmissibility factor TFE_x as follows:

$$\min_{\mathbf{x}_d} TFE_x(\mathbf{x}_d) \tag{6.4}$$

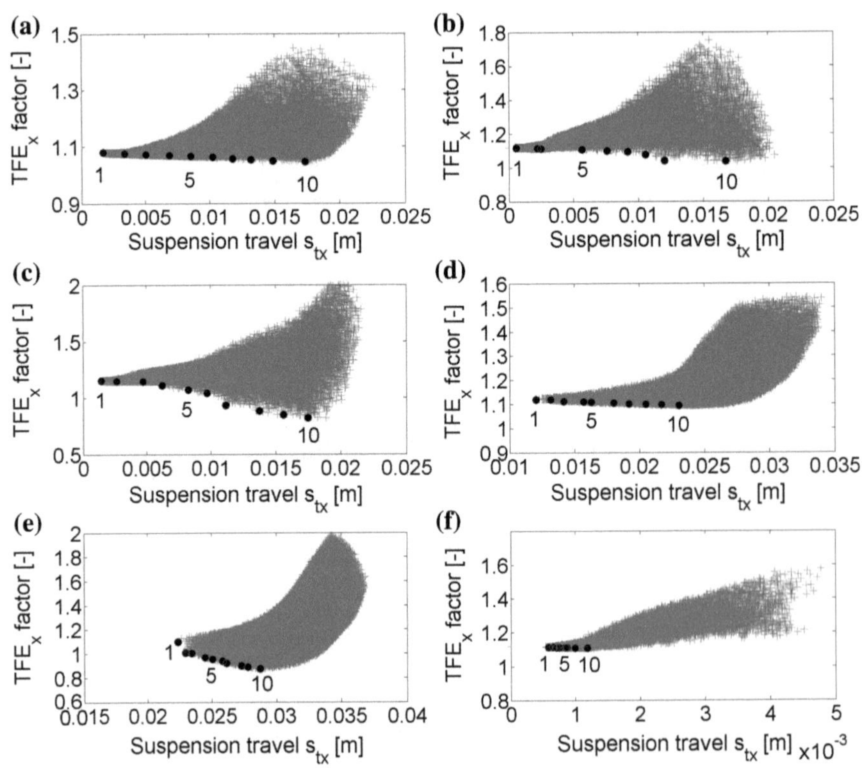

Fig. 6.15 Pareto-optimal point distribution (•) for the selected input vibration: AL1x (**a**), AT1x (**b**), AT2x (**c**), AT3x (**d**), AT4x (**e**), RV1x (**f**)

while the suspension travel is treated as the following inequality constraint:

$$s_{tx}(\mathbf{x}_d) \le s_{txj} \tag{6.5}$$

and the bounds of chosen decision variables are taken into account as:

$$(\mathbf{x}_d)_{min} \le \mathbf{x}_d \le (\mathbf{x}_d)_{max} \tag{6.6}$$

where s_{txj} is the constraint value imposed on suspension travel and $(\mathbf{x}_d)_{min}$ and $(\mathbf{x}_d)_{max}$ are the vectors including the minimum and maximum values of chosen design parameters.

The visco-elastic characteristics of horizontal seat suspension are sought for the selected input vibrations, ie. AL1x, AT1x, AT2x, AT3x, AT4x and RV1x. The minimisation of the constrained objective function (Eqs. 6.4–6.6) has allowed for discovering ten of the Pareto-optimal solutions that are illustrated in Fig. 6.15. For the first Pareto-optimal solution No. 1, the greatest limiting of the suspension travel s_{tx} is achieved, whereas the highest reduction of the transmissibility factor TFE_x is reached for the Pareto-optimal solution No. 10. The compromising solutions that efficiently decrease values both of the vibro-isolation criteria are designated as points No. 2–9.

The optimal force characteristics of principal visco-elastic elements contained in the horizontal seat suspension are presented in Fig. 6.16. The following characteristics represent the Pareto-optimal system configuration No. 10 for which the most

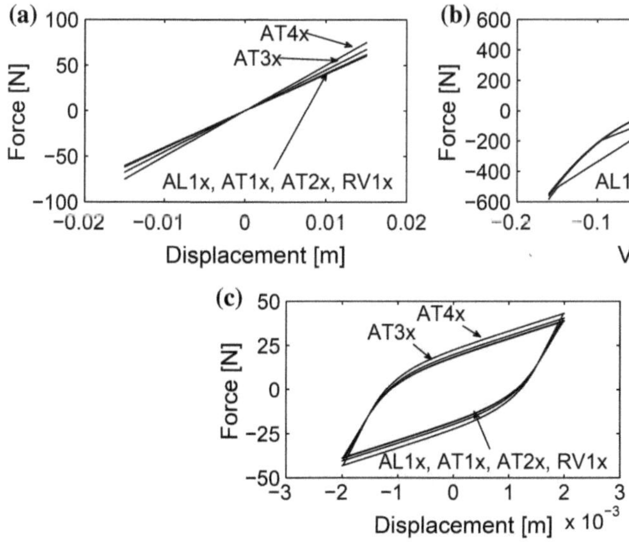

Fig. 6.16 Optimal force characteristics of the helical spring (**a**), hydraulic shock-absorber (**b**) and kinetic friction (**c**) for the selected input vibrations

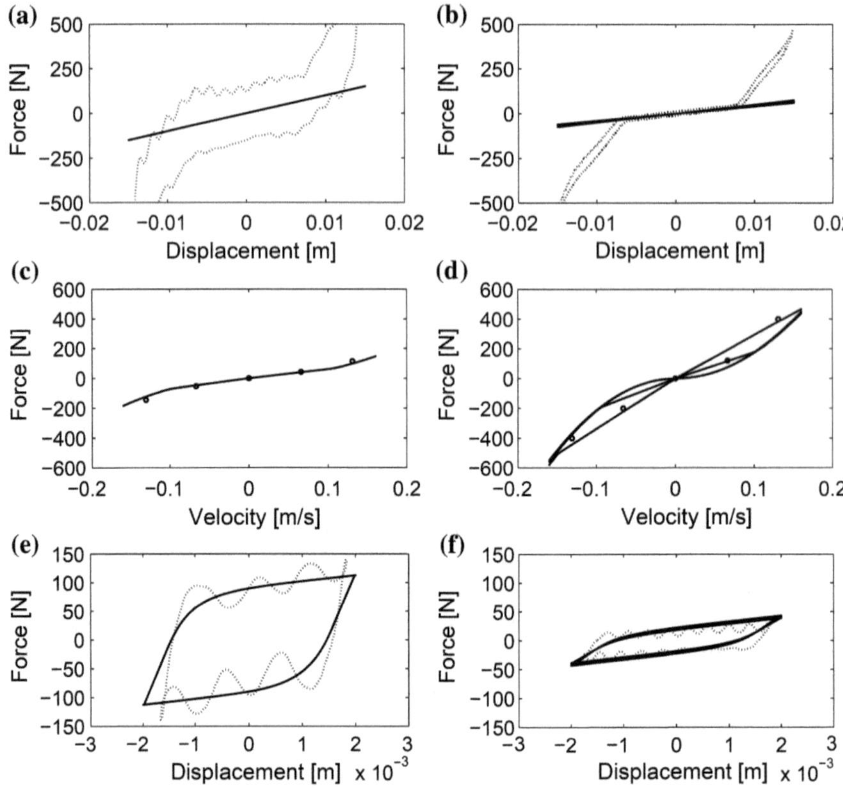

Fig. 6.17 Force characteristics of the conventional spring (**a**), hydraulic damper (**c**), kinetic friction (**e**) and force characteristics of the optimal spring (**c**), hydraulic damper (**d**), kinetic friction (**f**) achieved utilising simulation model (——), force–deflection measurement of the suspension system (······) and force–velocity measurement of the shock-absorber (○)

powerful vibration isolation of a driver is occurred. Although the suspension stroke has the largest values, an unwanted contact with the end-stop buffers appears only sometimes for the input vibration of medium amplitude range.

It should be noted that the visco-elastic characteristics illustrated in Fig. 6.16 show only little discrepancies in the case of considered excitation signals. In general, the desired system characteristics should possess the lower stiffness, the increased damping and the decreased suspension friction in relation to conventional system. For higher vibration amplitudes of the excitation signals AT3x and AT4x, rather negligible increase of stiffness and friction is noticeable, while the progressive damping tends towards a linear characteristic.

6.8.4 Experimental Verification of the System

For the purpose of experimental verification, the horizontal seat suspension is modified according to the optimisation results presented in the previous subsection. The stiffness of helical spring (Fig. 6.17a, b) is decreased by lowering the wire diameter, the viscous damping (Fig. 6.17c, d) is increased by using two shock-absorbers mounted in parallel, and the suspension friction (Fig. 6.17e, f) is reduced by applying needle bearings instead of commonly used plastic bearings.

In Fig. 6.18, the experimental set-up for testing the horizontal suspension with the seated human body is presented. The male subject is employed for the measurements and further analysis of the system vibro-isolation properties. The person being tested is placed on a seat cushion with a combined backrest support. The vertical suspension system is locked during the tests. Therefore, the seated occupant might be isolated against shocks and vibration exclusively by means of the horizontal seat suspension. The longitudinal random vibration along x-axis is generated during the measurements.

In Fig. 6.19, the transmissibility behaviour of the conventional and optimal horizontal seat suspension is shown for the excitation signals that represent real vibrations occurring in the cabins of various machines. As shown in this figure, the vibro-isolation properties of optimised system are considerably improved above the frequency of about 2 Hz. In the lower frequency range, an insignificant enlargement of the vibration amplitude is recognised as a result of the lower suspension stiffness that lowered the system resonance frequency.

The transmissibility factors TFE_x measured for the optimal suspension at different input signals are comparable with the results measured for the conventional system (Table 6.1). These measurement results are unsatisfactory due to the frequency

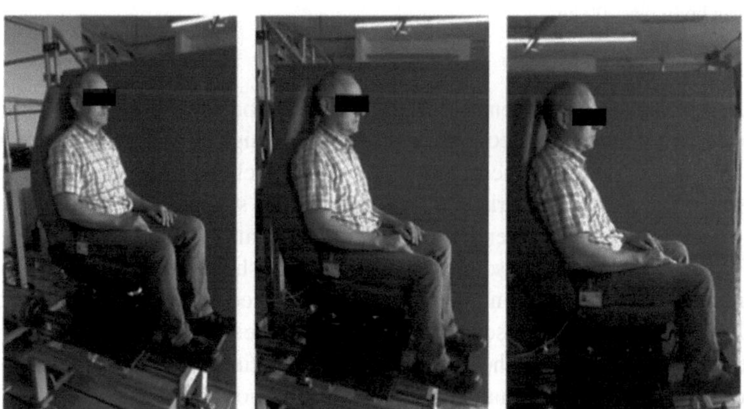

Fig. 6.18 Experimental set-up for evaluating the vibro-isolation properties of horizontal seat suspension with the seated human body

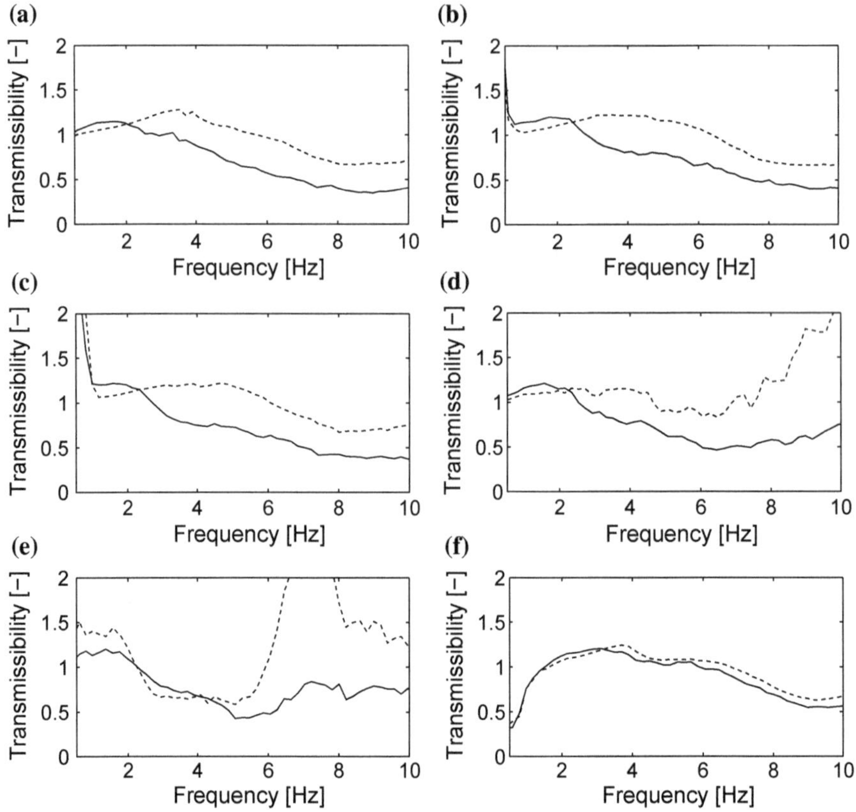

Fig. 6.19 Transmissibility functions of the conventional (----) and optimal (——) horizontal seat suspension for different spectral classes: AL1x (**a**), AT1x (**b**), AT2x (**c**), AT3x (**d**), AT4x (**e**), RV1x (**f**), human body mass 90 kg

weighting of acceleration signals, because the principal frequency weightings recommended for horizontal directions demonstrate the highest factors in the frequency range from 1 to 2 Hz. The measured suspension travels s_{tx} of optimal system have little larger values in comparison with conventional system; therefore, the higher suspension stroke is recommended for the softer vibration isolator.

In Fig. 6.20, the root mean square accelerations are shown for different exposure to vibration inputs. The vibration magnitudes are averaged over a measurement period of 300 s, and they are expressed as a function of the exposure duration. As follows from the presented graphs, the conventional and optimal systems give almost the same dynamic response of human body exposed to horizontal vibration. The daily exposure limit value ($1.15\ \mathrm{m/s^2}$) is exceeded for the excitation signal AT4x with the highest vibration amplitudes; however, the optimal seat suspension reduces the vibration exposure by a few per cent compared to the conventional system. In the case of all other signals, both systems operate similarly, while the levels of expo-

Table 6.1 Numerical values of the transmissibility factors and suspension travels obtained for excitation signals: AL1x, AT1x, AT2x, AT3x, AT4x and RV1x, human body mass 90 kg

Input vibration	Conventional		Optimal	
	TFE_x factor	s_{tx}, mm	TFE_x factor	s_{tx}, mm
AL1x	1.060	7.8	1.075	13.3
AT1x	1.104	1.0	1.111	8.5
AT2x	1.132	6.9	1.076	8.8
AT3x	1.090	22.1	1.135	22.9
AT4x	1.014	29.6	0.952	26.4
RV1x	1.107	0.2	1.118	1.7

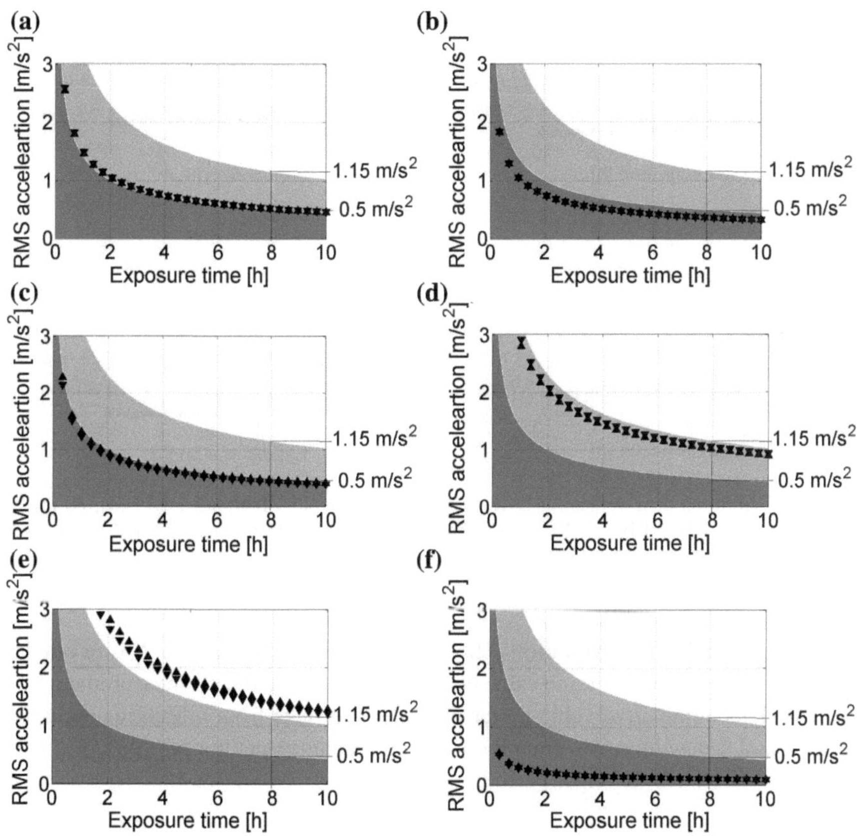

Fig. 6.20 Root mean square accelerations for different exposure to input vibrations: AL1x (**a**), AT1x (**b**), AT2x (**c**), AT3x (**d**), AT4x (**e**), RV1x (**f**) obtained using the conventional (▲) and optimal (▼) horizontal seat suspension, human body mass 90 kg

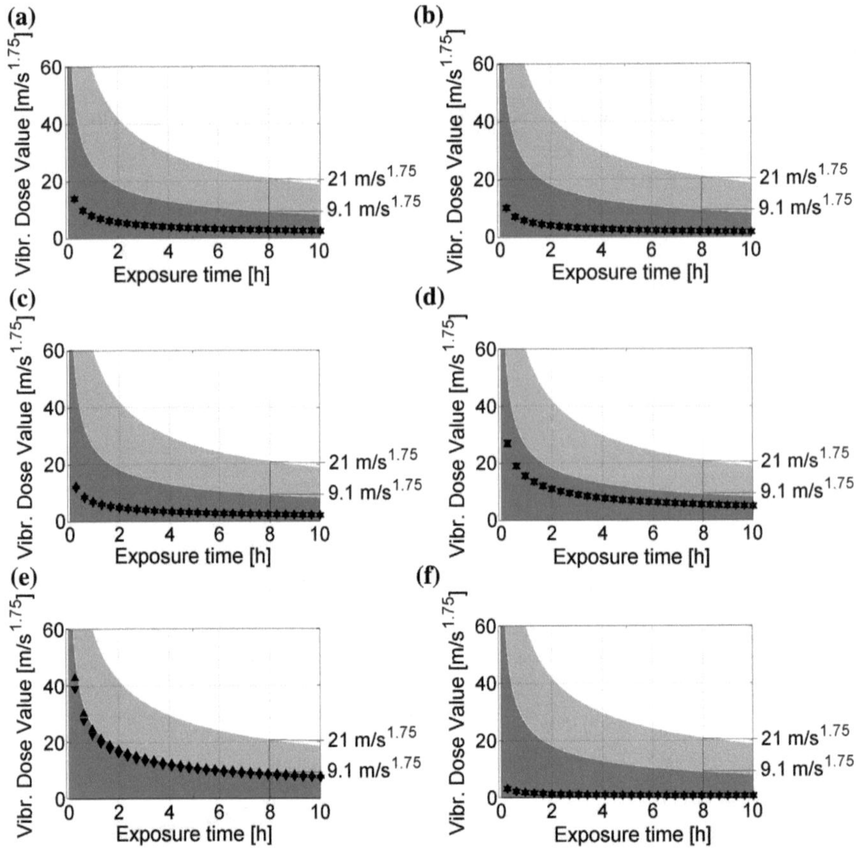

Fig. 6.21 Vibration dose values for different exposure to input vibrations: AL1x (**a**), AT1x (**b**), AT2x (**c**), AT3x (**d**), AT4x (**e**), RV1x (**f**) obtained using the conventional (▲) and optimal (▼) horizontal seat suspension, human body mass 90 kg

sure depend on the input vibration. The vibration dose values (Fig. 6.21) are rated as below the daily exposure action value (9.1 $m/s^{1.75}$) due to the short measurement time (300 s). This cumulative value better indicates the risks from vibrations including shocks, and it may increase when the working machine operator is exposed to horizontal vibration over a longer period.

6.9 Conclusions

In this chapter, a generalised methodology of shaping the vibro-isolation properties of passive vibration reduction systems is proposed. The visco-elastic system characteristics are selected, especially for strictly defined spectral classes of the input vibration that are produced employing innovative approach of the signal processing. A general simulation model of the passive suspension is created which allows to choose its non-linear characteristics by using the multi-criteria optimisation. An original minimisation procedure is developed in order to find Pareto-optimal system configuration with respect to the conflicted vibro-isolation criteria. The recommended methodology is inspected by means of experimental research that is performed on the exemplary horizontal seat suspension. The vibro-isolation properties of such a system are shaped for well-defined operating conditions.

References

1. British Standards Institution BS EN 13490. (2002). *Mechanical vibration—Industrial trucks—Laboratory evaluation and specification of operator seat vibration*, London.
2. International Organization for Standardization. (2003). *Agricultural wheeled tractors—Operator's seat—Laboratory measurement of transmitted vibration*, ISO 5007, Genewa.
3. International Organization for Standardization. (2000). *Earth-moving machinery–Laboratory evaluation of operator seat vibration*, ISO 7096, Genewa.
4. Bluthner, R., Seidel, H., & Hinz, B. (2008). Laboratory study as basis of the development for a seat testing procedure in horizontal directions. *International Journal of Industrial Ergonomics, 38*, 447–456.
5. Close, C. M., Frederick, D. K., & Newell, J. C. (2001). *Modeling and analysis of dynamic systems* (3rd ed.). New York: Wiley.
6. Rutzel, S., Hinz, B., & Wolfel, H. B. (2006). Modal description–A better way of characterizing human vibration behavior. *Journal of Sound and Vibration, 298*, 810–823.
7. Stein, G. J., Muka, P., Chmurny, R., Hinz, B., & Bluthner, R. (2007). Measurement and modelling of x-direction apparent mass of the seated human body–Cushioned seat system. *Journal of Biomechanics, 40*, 1493–1503.
8. Toward, M., & Griffin, J. (2011). The transmission of vertical vibration through seats: Influence of the characteristics of the human body. *Journal of Sound and Vibration, 330*, 6526–6543.
9. Davis, J. R., Johnson, R., Stepanek, J., & Fogarty, J. A. (2008). *Fundamentals of aerospace medicine*. London: Wolters Kluwer Health.
10. Bendat, J. S., & Piersol, A. G. (2004). *Methods for analysis and measurement of random signals (in Polish)*. Warsaw: Polish Scientific Publishers PWN.
11. Duke, M., & Goss, G. (2007). Investigation of tractor driver seat performance with non-linear stiffness and on-off damper. *Biosystems Engineering, 96*(4), 477–486.
12. International Organization for Standardization, *Mechanical vibration and shock—Evolution of human exposure to whole body vibration*, ISO 2631, Genewa 1997.
13. Griffin, M. J., Howarth, H. V. C., Pitts, P. M., Fischer, S., Kaulbars, U., Donati, P. M., et al. (2006). *Guide to good practice on whole-body vibration. Non-binding guide to good practice with a view to implementation of Directive 2002/44/EC on the minimum health and safety requirements regarding the exposure of workers to the risks arising from physical agents (vibrations)*. Luxembourg: European Commission.
14. Goh, E. G., & Noborio, K. (2014). Sensitivity analysis using sobol 'variance-based method on the haverkamp constitutive functions implemented in richards' water flow equation. *Malaysian Journal of Soil Science, 18*, 19–33.
15. Zhou, X., Lin, H., & Lin, H. (2008). *Global sensitivity analysis* (pp. 408–409). Encyclopedia of GIS: Springer.
16. Tarnowski, W. (2009). *Optimisation and polyoptimisation in mechatronics (in Polish)*. Koszalin: Academic Publishing of the Koszalin University of Technology.

17. Tarnowski, W., Krzyzynski, T., Maciejewski, I., & Olekiewicz, R. (2011). *Poly-optimization: A paradigm in engineering design in mechatronics* (pp. 141–156), Archive of Applied Mechanics 81, Springer.

Chapter 7
Control System Design

7.1 Nomenclature

$\mathbf{A}_i$	state (system) matrix ($i = x, y, z$)
$\mathbf{B}_{si}, \mathbf{B}_{ai}$	input matrices ($i = x, y, z$)
$\mathbf{C}_{1i}, \mathbf{C}_{2i}$	output matrices ($i = x, y, z$)
$\mathbf{D}_{1si}, \mathbf{D}_{1ai}$	feedthrough (feedforward) matrices
$\mathbf{D}_{2si}, \mathbf{D}_{2ai}$	($i = x, y, z$)
F_{ai}	applied force ($i = x, y, z$), N
$\mathbf{K}_i$	output feedback gain vector ($i = x, y, z$)
q_{1i}	displacement of the isolated body for selected direction of the vibration exposure ($i = x, y, z$), m
q_{si}	displacement of the input vibration ($i = x, y, z$), m
u_i	control signal ($i = x, y, z$), V
u_{pi}	predicted control signal ($i = x, y, z$), V
u_{ci}	constrained control signal ($i = x, y, z$), V
$\mathbf{w}_{si}$	vector of the external disturbances ($i = x, y, z$)
$\mathbf{x}_i$	state vector ($i = x, y, z$)
$\mathbf{y}_i$	measurement vector ($i = x, y, z$)
$\mathbf{z}_i$	output vector ($i = x, y, z$)

7.2 Introduction

There are several types of actuators, e.g. hydraulic, pneumatic, electromagnetic, that are often used in active vibration reduction systems. Selecting the type of actuator for a specific application depends on the system requirements. For example, vehicle and seat suspensions require a high-static stiffness; therefore, an application of the hydraulic actuators seems to be an appropriate solution [1, 2]. Although the pneumatic actuators demonstrate lower static stiffness [3], they are of high interest due to

T. Krzyzynski et al., *Modelling and Control Design of Vibration Reduction Systems*, https://doi.org/10.1007/978-3-030-03047-6_7

the fact that the compressed air is generally available in the practical operation [4]. A significant part of the active vibration isolators is the electromagnetic systems because of a large number of commercially available actuators. Electromagnetic drives are fast enough for automotive suspensions; however, they have limited opportunities while working at relatively high vibration amplitudes [5]. In this chapter, the MR damper and pneumatic artificial muscles are used for controlling vibrations of the horizontal seat suspension.

7.3 System Structure

The system structure for designing the advanced control algorithm of semi-active and active vibration isolators is presented in Fig. 7.1. The developed control system consists of the primary controller that calculates the desired active force. Such a force should be reproduced in the vibration reduction system by using the control element. The secondary controller is employed to calculate the control signal based on an inverse model of the specific control element applied in the system. The predictor PD speeds up the semi-active or active system control because the actuating time of the control element can be partly eliminated. The proposed structure assists in finding a suitable control algorithm, which is able to predict the desired active force and adapt the vibration reduction system to various working conditions.

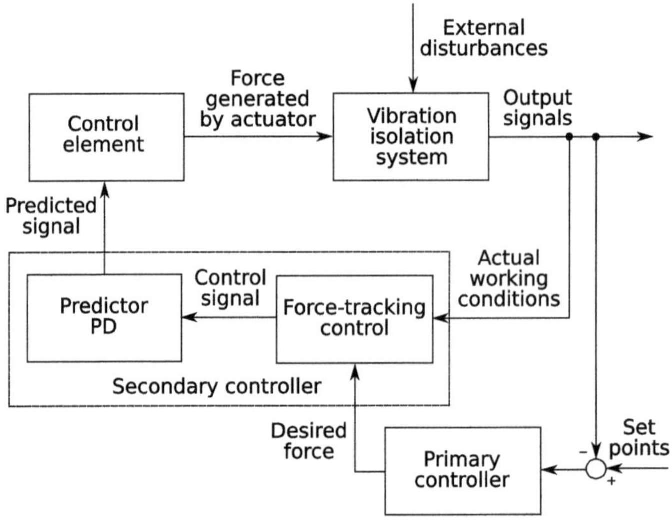

Fig. 7.1 Block diagram of the control structure of vibration reduction systems

7.4 Primary Controller

An algorithm of the primary controller defines the desired active force F_{ai} that should be generated in the vibration reduction system in order to compensate the kinematic excitations q_{si} (for directions of the vibration exposure x, y or z). The most popular methods of control synthesis are [6]: the Ziegler–Nichols techniques of tuning a PID controller, model-based design and frequency control approaches. All of these methods are suitable only for the control design of a single-input, single-output system. The methods based on the optimal control theory [7] are often used for the control synthesis of linear systems with multiple inputs and multiple outputs: LQ control (linear-quadratic) or LQG control (linear-quadratic-Gaussian).

The most widespread methods for nonlinear control are [8]: the phase-plane method, the method of Popov and the gain scheduling techniques. Such methods are only useful for controlling the system dynamics that is described by the models with fixed parameters. In the existing literature [9, 10], the adaptive methods are extensively used that employ the linear object model, but the system parameters change with their state (linear parameter-varying control). The model is utilised to simulate the dynamics of nonlinear vibration reduction system by identifying its parameters in real time and tuning controller based on the set of evaluated model parameters.

In this chapter, a structure of the primary controller is determined on the basis of the simplified model of vibration reduction system (Fig. 7.2). Such a model describes the most important dynamic properties of many vibration isolation systems often discussed in the literature [11–13]. It employs a linear model of the suspension system for the purpose of a control synthesis using classical methods.

The state-space model of the hybrid vibration reduction system (Fig. 7.3) can be obtained by using the linear fractional transformation (LFT) technique [14], which is usually used to determine the incorporation of a controller into a system by grouping signals as the sets of external inputs $\mathbf{w}_i(t)$ and outputs $\mathbf{z}_i(t)$, inputs to the controller $\mathbf{y}_i(t)$ and an output from the controller $F_{ai}(t)$.

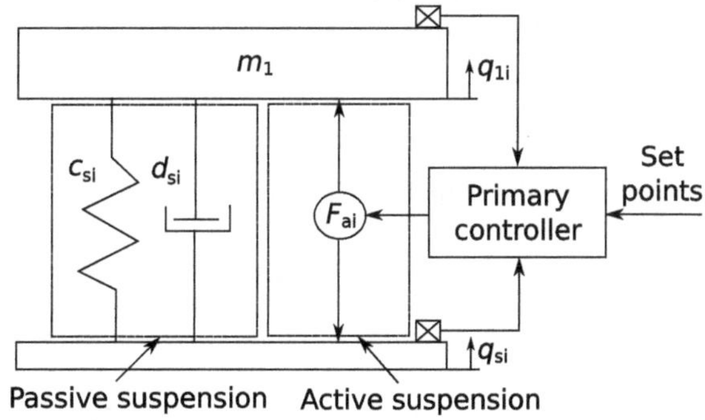

Fig. 7.2 Simplified model of the hybrid vibration reduction system for one direction of the vibration transmission (i = x, y, z)

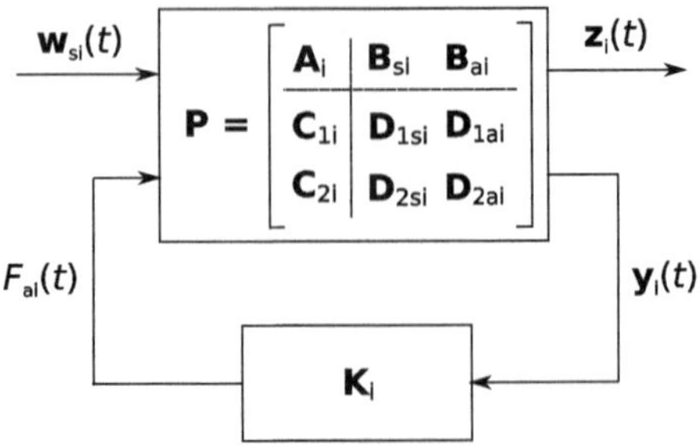

Fig. 7.3 Linear fractional transformation (LFT)

The state variables which allow to shape vibro-isolation properties of the system are chosen as follows:

$$\mathbf{x}_i(t) = [q_{1i} - q_{si}, \dot{q}_{1i}]^T, \qquad i = x, y, z \tag{7.1}$$

where $q_{1i} - q_{si}$ is the relative displacement of vibration reduction system, and $\dot{q}_{1i}$ is the absolute velocity of the isolated body. The absolute displacement q_{si} and the absolute velocity $\dot{q}_{si}$ are used as disturbances caused by road roughness:

$$\mathbf{w}_i(t) = [q_{si}, \dot{q}_{si}]^T, \qquad i = x, y, z \tag{7.2}$$

Then, the state-space equation of hybrid vibration reduction system is presented as:

$$\dot{\mathbf{x}}_i(t) = \mathbf{A}_i\mathbf{x}_i(t) + \mathbf{B}_{si}\mathbf{w}_i(t) + \mathbf{B}_{ai}F_{ai}(t) \tag{7.3}$$

where

$$\mathbf{A}_i = \begin{bmatrix} 0 & 1 \\ -\frac{c_{si}}{m_1} & -\frac{d_{si}}{m_1} \end{bmatrix}, \quad \mathbf{B}_{si} = \begin{bmatrix} 0 & -1 \\ 0 & \frac{d_{si}}{m_1} \end{bmatrix}, \quad \mathbf{B}_{ai} = \begin{bmatrix} 0 \\ \frac{1}{m_1} \end{bmatrix} \tag{7.4}$$

In order to satisfy the performance requirement, the acceleration of the isolated body $\ddot{q}_{1i}$ and the relative displacement of vibration reduction system $q_{1i} - q_{si}$ are defined as the controlled outputs:

$$\mathbf{z}_i(t) = [\ddot{q}_{1i}, q_{1i} - q_{si}]^T, \qquad i = x, y, z \tag{7.5}$$

therefore, the output equation reads:

$$\mathbf{z}_i(t) = \mathbf{C}_{1i}\mathbf{x}_i(t) + \mathbf{D}_{1si}\mathbf{w}_i(t) + \mathbf{D}_{1ai}F_{ai}(t) \tag{7.6}$$

where

$$\mathbf{C}_{1i} = \begin{bmatrix} -\frac{c_{si}}{m_1} & -\frac{d_{si}}{m_1} \\ 1 & 0 \end{bmatrix}, \quad \mathbf{D}_{1si} = \begin{bmatrix} 0 & \frac{d_{si}}{m_1} \\ 0 & 0 \end{bmatrix}, \quad \mathbf{D}_{1ai} = \begin{bmatrix} \frac{1}{m_1} \\ 0 \end{bmatrix} \tag{7.7}$$

If the relative displacement of vibration reduction system $q_{1i} - q_{si}$ and the absolute velocity of the isolated body $\dot{q}_{1i}$ are measurable, then the following vector can be expressed as:

$$\mathbf{y}_i(t) = [q_{1i} - q_{si}, \dot{q}_{1i}]^T, \qquad i = x, y, z \tag{7.8}$$

and the measurement equation reads:

$$\mathbf{y}_i(t) = \mathbf{C}_{2i}\mathbf{x}_i(t) + \mathbf{D}_{2si}\mathbf{w}_i(t) + \mathbf{D}_{2ai}F_{ai}(t) \tag{7.9}$$

where

$$\mathbf{C}_{2i} = \begin{bmatrix} 1 & 0 \\ 0 & 1 \end{bmatrix}, \ \mathbf{D}_{si} = \begin{bmatrix} 0 & 0 \\ 0 & 0 \end{bmatrix}, \ \mathbf{D}_{ai} = \begin{bmatrix} 0 \\ 0 \end{bmatrix} \tag{7.10}$$

In practice, the relative displacement of seat suspension is measured by a cable extension position sensor. The absolute velocity of the suspended body is typically evaluated by integrating and filtering the acceleration signal which is measured by a linear accelerometer.

A controller is determined by formulating the state feedback control problem in the following form:

$$F_{ai}(t) = \mathbf{K}_i\mathbf{y}_i(t) = \mathbf{K}_i\mathbf{C}_{2i}\mathbf{x}_i(t) \tag{7.11}$$

wherein $\mathbf{K}_i$ is the output feedback gain vector calculated as:

$$\mathbf{K}_i = \begin{bmatrix} k_{1i} \ k_{2i} \end{bmatrix} \tag{7.12}$$

where k_{1i} is the proportionality factor of the relative displacement feedback loop, and k_{2i} is the proportionality factor of the absolute velocity feedback loop.

7.5 Secondary Controller

The calculated force F_{ai} should be reproduced in the vibration reduction system by using controllable spring, controllable damper or the force actuator. Typically, the internal force feedback is used to realise the desired force F_{ai} for the specific direction of vibration exposure ($i = x, y, z$) [15]. In this case, the controller is designed to track the force trajectory using an additional force transducer that is able to measure the actual force F_{mi} generated by controllable drive. This type of control is shown in Fig. 7.4.

In this book, it is proposed to reproduce the desired force F_{ai} without the necessity of applying any additional force feedback. Such a formulated control strategy requires the application of the secondary controller that calculates the force generated by the control element on the basis of its inverse model. Actual value of the control signal

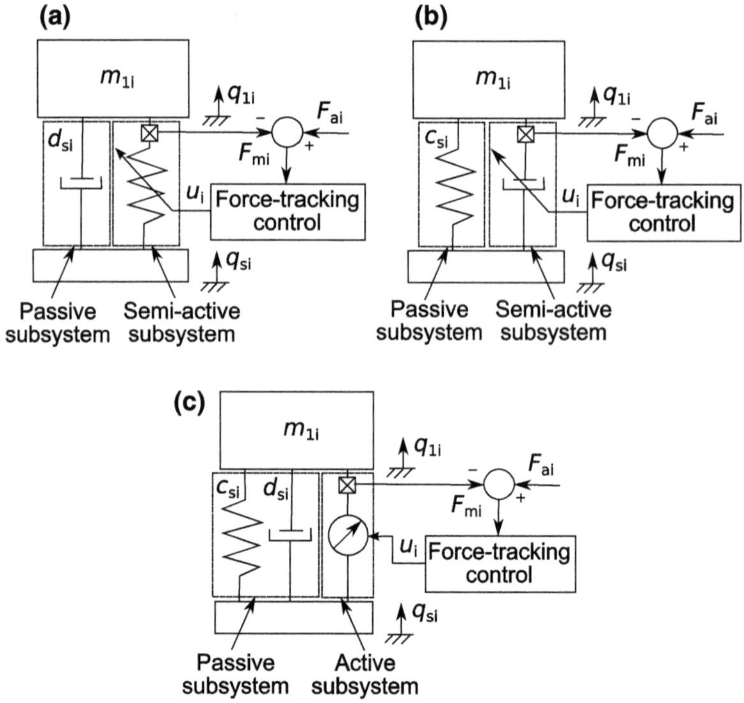

Fig. 7.4 Simplified models of the vibration isolation systems: semi-active system with spring control (**a**), semi-active system with damper control (**b**) and active with force actuator (**c**) for one direction of the vibration transmission (i = x, y, z)

u_i should be calculated for the specific type of control element (adjustable spring, adjustable damper or force actuator) according to the following functions:

- adjustable spring

$$u_i = \begin{cases} f(q_{1i} - q_{si}, F_{ai}) & \text{for } F_{ai}(q_{1i} - q_{si}) > 0 \\ 0 & \text{for } F_{ai}(q_{1i} - q_{si}) \leq 0 \end{cases} \tag{7.13}$$

- adjustable damper

$$u_i = \begin{cases} f(\dot{q}_{1i} - \dot{q}_{si}, F_{ai}) & \text{for } F_{ai}(\dot{q}_{1i} - \dot{q}_{si}) > 0 \\ 0 & \text{for } F_{ai}(\dot{q}_{1i} - \dot{q}_{si}) \leq 0 \end{cases} \tag{7.14}$$

- force actuator

$$u_{\mathrm{i}} = f(q_{1\mathrm{i}} - q_{\mathrm{si}}, \dot{q}_{1\mathrm{i}} - \dot{q}_{\mathrm{si}}, F_{\mathrm{ai}}) \tag{7.15}$$

where F_{ai} is the desired force in the system used for the specific direction of vibration exposure ($\mathrm{i} = x, y, z$), $q_{1\mathrm{i}} - q_{\mathrm{si}}$ and $\dot{q}_{1\mathrm{i}} - \dot{q}_{\mathrm{si}}$ are the actual values of relative displacement and relative velocity of the control elements.

In order to speed up the force-tracking control, the predictor PD (proportional–derivative) is employed because its dynamic properties overcome the delays occurring in the system [16, 17]. The mathematical model of such a predictor can be described using the following equation:

$$u_{\mathrm{pi}} = t_{\mathrm{ci}}\dot{u}_{\mathrm{i}} + u_{\mathrm{i}}, \qquad \mathrm{i} = x, y, z \tag{7.16}$$

where u_{i} is the control signal calculated on the basis of the inverse model (input to predictor PD), and t_{ci} is the actuating time of the control element.

Finally, the predicted signal u_{pi} must be limited to a specified interval before forwarding it to the control element. Even if the desired force F_{ai} is calculated as a very high value, then the signal sent to the control element is constrained within the following range:

$$u_{\mathrm{ci}} = \begin{cases} u_{\mathrm{min}} & \text{for} \quad u_{\mathrm{pi}} < u_{\mathrm{min}} \\ u_{\mathrm{pi}} & \text{for} \quad u_{\mathrm{min}} \leq u_{\mathrm{pi}} < u_{\mathrm{max}}, \\ u_{\mathrm{max}} & \text{for} \quad u_{\mathrm{pi}} \geq u_{\mathrm{max}} \end{cases} \qquad \mathrm{i} = x, y, z \tag{7.17}$$

where u_{min} and u_{max} are the minimum and maximum values of the signal that controls the adjustable spring, adjustable damper or force actuator.

7.6 Example: Control System of a Horizontal Seat Suspension with the MR Damper

7.6.1 Model of the Semi-active Suspension

The physical model of a horizontal seat suspension with the MR damper is shown in Fig. 7.5. The electric power control is applied by means of the programmable power supply that regulates the current flowing through the coil windings. In such a system, the MR fluid inside a damper changes its properties (mainly its viscosity) with the application of a magnetic field.

The equation of motion takes a similar form to the model of passive suspension system that has been discussed in Sect. 3.5:

$$m_1\ddot{q}_{1\mathrm{x}} = -2F_{\mathrm{cx1}} - F_{\mathrm{cx2}} - F_{\mathrm{dx2}} - F_{\mathrm{ax1}} \tag{7.18}$$

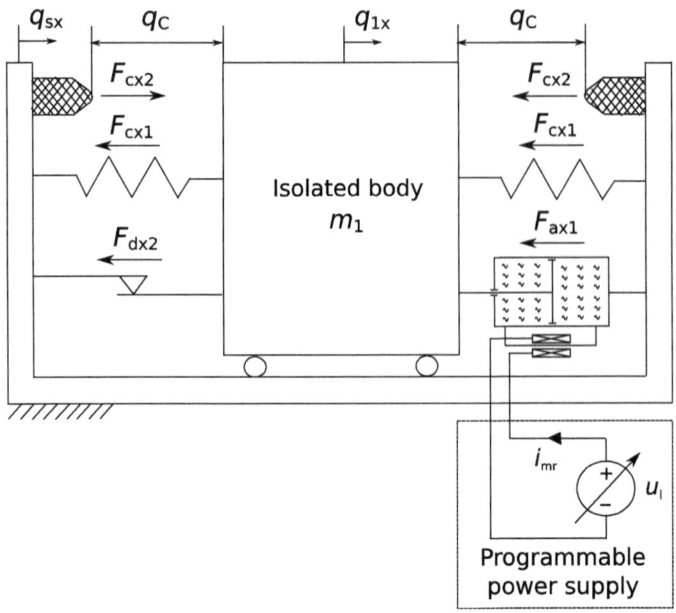

Fig. 7.5 Physical model of a horizontal seat suspension with the MR dumper

where m_1 is the suspended mass, and q_{1x} is the horizontal displacement of suspended body. The fundamental forces acting in the suspension system, i.e. the mechanical spring force F_{cx1}, the forces from end-stop buffers F_{cx2} and the overall friction force F_{dx2}, have been presented in Sect. 3.4. The hydraulic shock absorber is replaced with the MR damper device; therefore, the semi-active force F_{ax1} can be introduced into the suspension system. The mathematical model of the MR damper has been also presented in the same section.

7.6.2 Control System Synthesis of the Semi-active Suspension

The developed control algorithm mimics the action of a damper suspended in the inertial reference point (skyhook damper algorithm [18]). Therefore, according to Eq. (7.11), the primary controller uses only one feedback loop ($k_{1x} = 0$) from the absolute velocity $\dot{q}_{1x}$ of the suspended body:

$$F_{ax} = k_{2x}\dot{q}_{1x} \tag{7.19}$$

where k_{2x} is the primary controller setting.

If the MR damper model is determined (see Sect. 3.4.5), the desired force F_{ax} can be obtained partially by injecting an appropriate current i_{mr} into the coil windings in accordance with the actual piston displacement $q_{1x} - q_{sx}$ and velocity $\dot{q}_{1x} - \dot{q}_{sx}$ of the damper. The controlled current is calculated from Eqs. (3.34)–(3.37) using the following relations:

$$i_{mr} = \frac{-a_1 (\dot{q}_{1x} - \dot{q}_{sx}) - b_1 z_{mr} + \operatorname{sign}(z_{mr}) \sqrt{\Delta}}{2 b_2 z_{mr}} \tag{7.20}$$

with a function Δ that is calculated as:

$$\begin{aligned} \Delta = & (a_1 (\dot{q}_{1x} - \dot{q}_{sx}) + b_1 z_{mr})^2 + \\ & -4 b_2 z_{mr} (a_0 (\dot{q}_{1x} - \dot{q}_{sx}) + c_{mr} (q_{1x} - q_{sx}) + b_0 z_{mr} + f_{mr} - F_{ax}) \end{aligned} \tag{7.21}$$

where F_{ax} is the desired MR damper force, c_{mr} is the stiffness coefficient, z_{mr} is the hysteretic variable given by Eq. (3.35), f_{mr} is the force offset, and a_1, a_0 and b_2, b_1, b_0 are the polynomial coefficients expressed with respect to the input current (Eqs. 3.36–3.37). These coefficients are evaluated by means of additional MR damper measurements, and results of the viscous damping d_{mr} and scale factor of the hysteresis α_{mr} are shown in Fig. 7.6.

In Fig. 7.7, the graphical representation of the MR damper inverse model is shown. It should be noted that the MR damper is a semi-active device and the desired force F_{ax} can be realised only if this force and the damper velocity have the same sign. Then, the calculated input signal of the MR damper varies in the range of 0 A (minimum value) and 4 A (maximum value) and depends on the actual value of the desired MR damper force and its actual velocity. If the desired force and damper velocity have a sign opposite to each other, then the input signal is settled to zero.

In the following book, it is assumed that the controlled current i_{mr} is proportional to the input signal u_i using the programmable power supply. In order to eliminate the actuating time t_{mr} of MR damper, the voltage input signal is calculated by the predictor PD as follows:

$$u_{mr} = \frac{1}{k_{mr}} \left(t_{mr} \dot{i}_{mr} + i_{mr} \right) \tag{7.22}$$

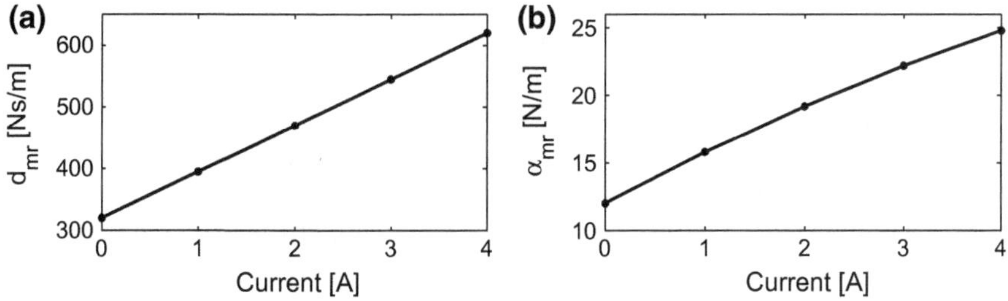

Fig. 7.6 Measured (—) and approximated (•) viscous damping coefficient d_{mr} and scale factor of the hysteresis α_{mr} as a function of the input current

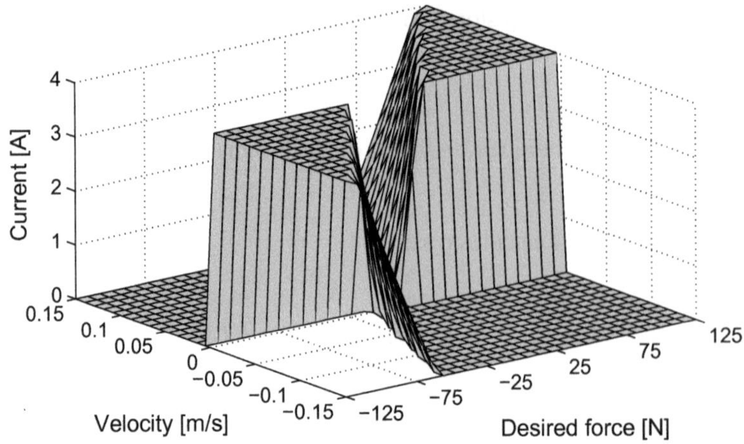

Fig. 7.7 Inverse model $i_{\text{mr}} = f(F_{\text{ax}}, \dot{q}_{1\text{x}} - \dot{q}_{\text{sx}})$ of the MR damper at different values of the desired force and velocity

where k_{mr} is the static gain of MR damper, and u_{i} is the control signal. The electric input signal for MR damper is then limited to the operating range:

$$u_{\text{c}} = \begin{cases} u_{\min} \text{ for } & u_{\text{mr}} < u_{\min} \\ u_{\text{mr}} \text{ for } & u_{\min} \le u_{\text{mr}} < u_{\max} \\ u_{\max} \text{ for } & u_{\text{mr}} \ge u_{\max} \end{cases} \tag{7.23}$$

where $u_{\min}$ is the minimum input voltage that is responsible for the lowest damping force, and $u_{\max}$ is the maximum input voltage that is required to generate the highest damping force.

7.6.3 Semi-active System Verification

The unknown model parameters are determined by using force–velocity measurements of the MR damper. The sinusoidal cycling of the MR damper is applied with an amplitude of 12.5 mm and a frequency of 1.8 Hz. The tests are performed at constant current flowing through the coil windings, i.e. 0, 2 and 4 A. The measured force in velocity domain is fitted to the simulation results (Fig. 7.8) by appropriate selection of the model parameters. Parameter values used by the MR damper model are shown in Table 7.1.

The system verification is performed by using the electrohydraulic shaker shown in Fig. 7.9. The system dynamics of semi-active horizontal suspension is investigated for excitation signals at various excitation intensities in the range of frequency 1–10 Hz, i.e. WN1x ($1.02\ \text{m/s}^2$), WN2x ($1.36\ \text{m/s}^2$) and WN3x ($1.91\ \text{m/s}^2$).

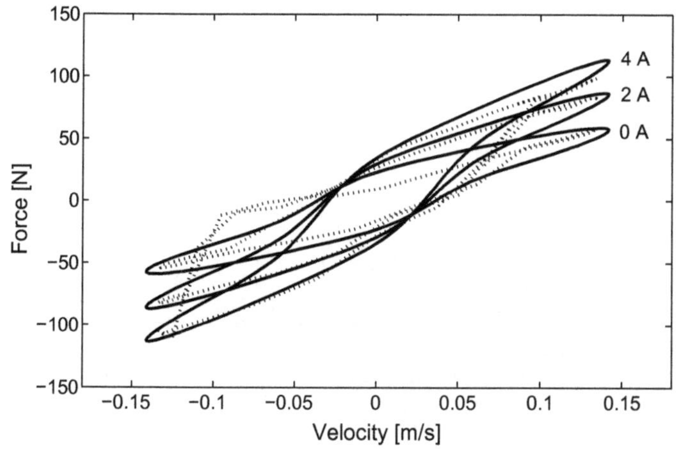

Fig. 7.8 Force characteristics of the MR damper at constant values of the current flowing through the coil windings obtained using simulation model (—) and measurement (''''''')

Table 7.1 Parameter values used by the MR damper model

System component	Parameter	Value	Unit
MR damper	Stiffness coefficient (c_{mr})	1000	N/m
	Force offset (f_{mr})	0	N
	First parameter shaping force hysteresis of the MR damper (β_{mr})	40	s/m
	Second parameter shaping force hysteresis of the MR damper (δ_{mr})	1.4	1/m
	First polynomial coefficient approximating viscous damping of the MR damper (a_0)	320	Ns/m
	Second polynomial coefficient approximating viscous damping of the MR damper (a_1)	75	Ns/(Am)
	First polynomial coefficient approximating scale factor of the hysteresis (b_0)	12	N
	Second polynomial coefficient approximating scale factor of the hysteresis (b_1)	4	N/A
	Third polynomial coefficient approximating scale factor of the hysteresis (b_2)	−0.2	N/A^2
Programmable power supply	Actuating time (t_{mr})	0.01	s
	Static gain (k_{mr})	1	A/V

The power spectral densities and transmissibility functions of the horizontal semi-active suspension loaded by different dummy masses (30, 55 and 80 kg) are shown in Figs. 7.10, 7.11 and 7.12. The simulation and measurement results are close to each other; therefore, the model assumptions and simplifications are correctly formulated in this book. Moreover, the transmissibility functions are lower than 1 over the

Fig. 7.9 Experimental set-up for the system verification of the semi-active horizontal seat suspension

considered frequency range of random excitation signal. For this reason, the control system of semi-active seat suspension seems to be correctly formulated.

In Table 7.2, simulated and measured transmissibility factors, suspension travels and relative errors of the semi-active horizontal seat suspension are presented. As shown in this table, the maximum error does not exceed 8%; therefore, it proves a correct modelling of the seat suspension system aided by the proposed semi-active vibration control.

7.7 Example: Control System of a Horizontal Seat Suspension with the Pneumatic Muscles

7.7.1 Model of the Active Suspension

The physical model of a horizontal seat suspension with the pneumatic muscles is shown in Fig. 7.13. The air-pressure control is applied by means of the proportional pressure control valves that regulates the inflating/exhausting of the pneumatic muscles located on both sides of the isolated body. In such a system, the pressure in the pneumatic muscles can be changed very fast, and therefore, an active force for the horizontal seat suspension is provided.

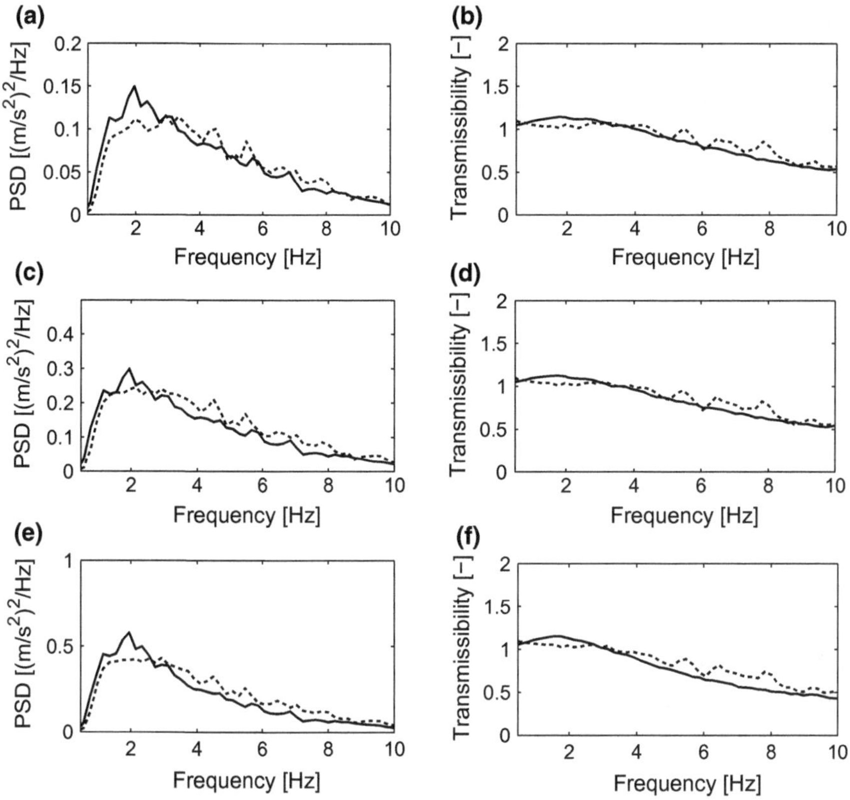

Fig. 7.10 Simulated (——) and measured (----) power spectral densities and transmissibility functions of the semi-active horizontal seat suspension for different spectral classes: WN1x (**a**, **b**), WN2x (**c**, **d**), WN3x (**e**, **f**), mass load 30 kg

The equation of motion takes a similar form as in the case of the passive suspension model that is shown in Sect. 3.5 as follows:

$$m_1\ddot{q}_{1x} = -2F_{cx1} - F_{cx2} - F_{dx1} - F_{dx2} + F_{ax1} - F_{ax2} \tag{7.24}$$

where m_1 is the mass of isolated body, and q_{1x} is the horizontal displacement of the isolated body. The mathematical models of the basic forces in the system, i.e. the mechanical spring force F_{cx1}, the forces from end-stop buffers F_{cx2} limiting the maximum suspension travel, the force of hydraulic damper F_{dx1} and the friction force F_{dx2} of suspension mechanism, have been exhaustively discussed in Sect. 3.4. However, the desired active force F_{ax} is introduced into the suspension system by using two pneumatic muscles that work in opposite directions. The mathematical model of the pneumatic muscle has been also presented in the same section.

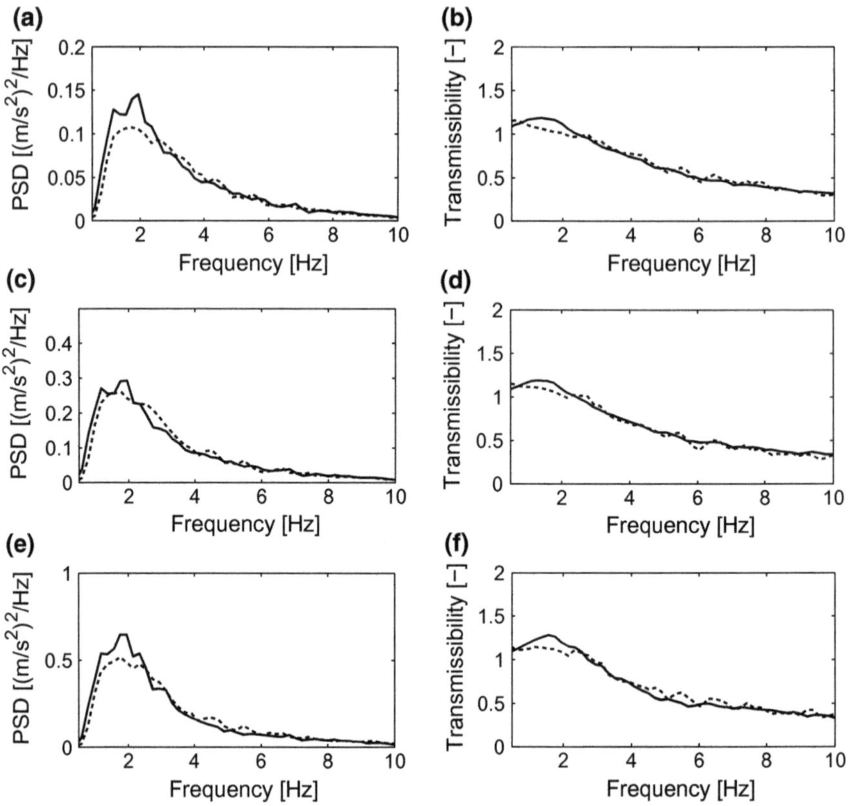

Fig. 7.11 Simulated (——) and measured (----) power spectral densities and transmissibility functions of the semi-active horizontal seat suspension for different spectral classes: WN1x (**a**, **b**), WN2x (**c**, **d**), WN3x (**e**, **f**), mass load 55 kg

7.7.2 Control System Synthesis of the Active Suspension

According to Eq. (7.11), the primary controller is determined by formulating the state feedback control problem in the following form:

$$F_{ax} = k_{1x}\,(q_{1x} - q_{sx}) + k_{2x}\dot{q}_{1x} \tag{7.25}$$

where F_{ax} is the desired active that should be generated for the longitudinal direction of vibration exposure, $q_{1x} - q_{sx}$ is the relative displacement of suspension system, $\dot{q}_{1x}$ is the absolute velocity of isolated body, and k_{1x} and k_{2x} are the output feedback gains.

Since the desired active force F_{ax} must be generated when the isolated body moves forward and backward, two separate pneumatic muscles are required. The first pneumatic muscle is utilised to apply the force F_{ax1} that compensates forward

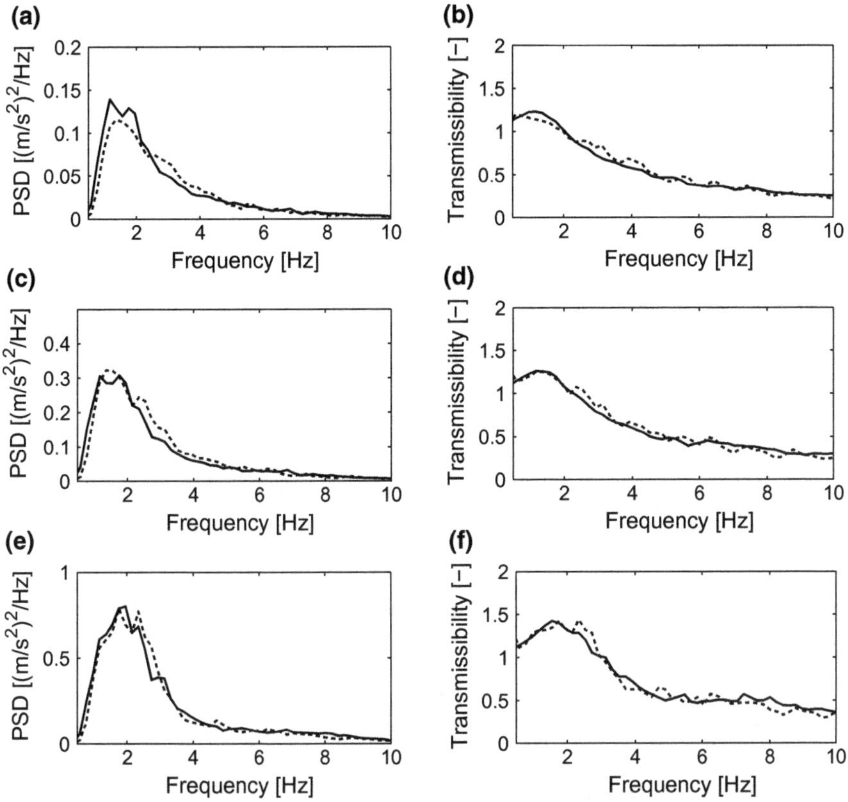

Fig. 7.12 Simulated (—) and measured (----) power spectral densities and transmissibility functions of the semi-active horizontal seat suspension for different spectral classes: WN1x (**a**, **b**), WN2x (**c**, **d**), WN3x (**e**, **f**), mass load 80 kg

Table 7.2 Simulated and measured transmissibility factors, suspension travels and relative errors of the semi-active horizontal seat suspension for excitation signals: WN1x, WN2x and WN3x

Suspended mass	Input vibration	Simulation		Measurement		Relative error	
		TFE_x factor	s_{tx} (mm)	TFE_x factor	s_{tx} (mm)	δ_{TFE_x} (%)	$\delta_{s_{tx}}$ (%)
30 kg	WN1x	1.053	15.4	1.006	14.3	4.7	7.6
	WN2x	1.034	23.6	0.989	22.9	4.5	3.1
	WN3x	1.023	30.3	0.983	31.5	4.1	3.9
55 kg	WN1x	0.982	24.8	0.922	23.5	6.5	5.5
	WN2x	0.983	34.3	0.931	36.0	5.5	4.7
	WN3x	1.039	38.6	0.971	41.1	7.0	6.1
80 kg	WN1x	0.925	31.7	0.892	30.3	3.6	4.6
	WN2x	0.979	38.7	0.969	41.7	1.0	7.2
	WN3x	1.121	45.8	1.132	46.8	0.8	2.2

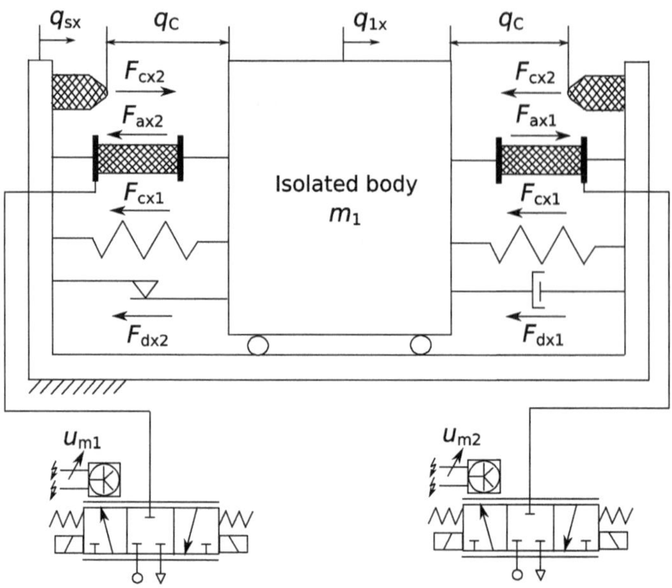

Fig. 7.13 Physical model of a horizontal seat suspension with the pneumatic muscles

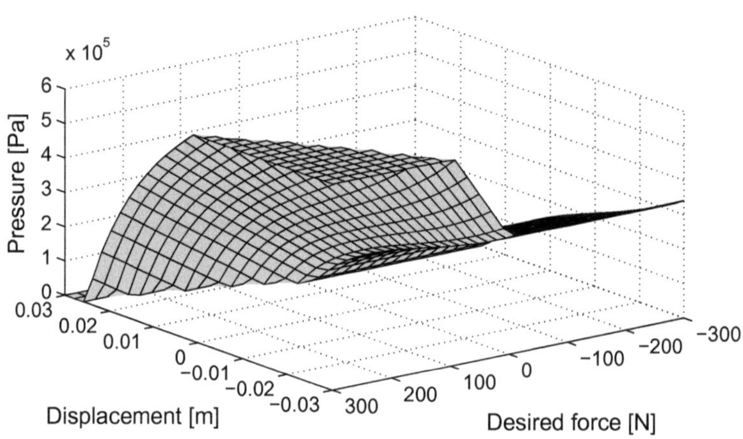

Fig. 7.14 Inverse model $p_{m1} = f(F_{ax1}, q_{1x} - q_{sx})$ of the right muscle at different values of the desired force and suspension travel

vibrations of the isolated body. The second muscle creates the force F_{ax2} that is used to actively reduce vibrations in the backward direction. The following discontinuous arrangement is employed to divide the desired active force:

$$F_{ax1} = \begin{cases} F_{ax} & \text{for} \quad F_{ax} > 0 \\ 0 & \text{for} \quad F_{ax} \leq 0 \end{cases} \qquad F_{ax2} = \begin{cases} 0 & \text{for} \quad F_{ax} \geq 0 \\ -F_{ax} & \text{for} \quad F_{ax} < 0 \end{cases} \tag{7.26}$$

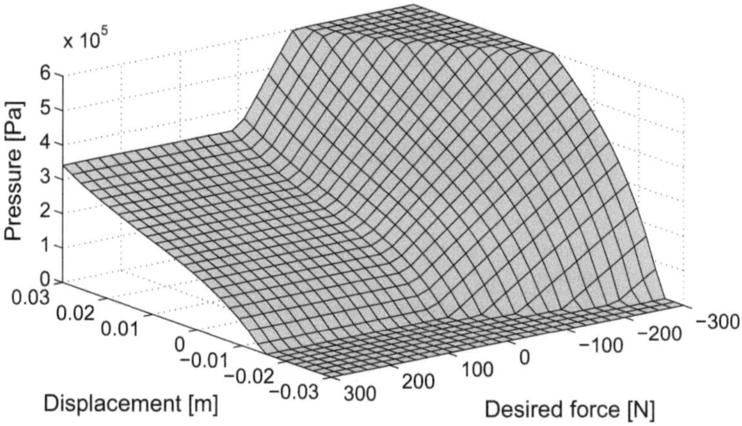

Fig. 7.15 Inverse model $p_{m2} = f(F_{ax2}, q_{1x} - q_{sx})$ of the left muscle at different values of the desired force and suspension travel

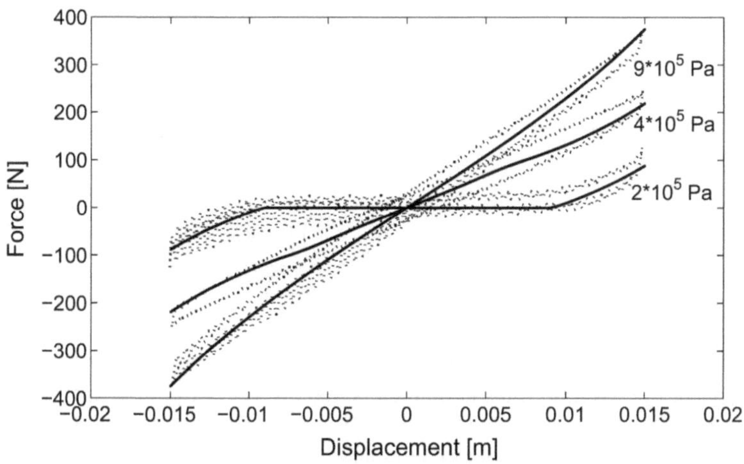

Fig. 7.16 Static force of the suspension system at constant values of the air-pressure inside the muscles obtained using simulation model (—) and measurement (⋯)

where F_{ax1} and F_{ax2} are the forces to be realised by the pneumatic muscles placed on the right and left sides of the isolated body, respectively.

The forces F_{ax1} and F_{ax2} can be reproduced by obtaining an appropriate pressures p_{m1} and p_{m2} inside the particular muscles in accordance with the measured actual relative displacement $q_{1i} - q_{si}$ of the suspension system. Therefore, the air-pressure inside the pneumatic muscles is calculated from Eqs. (3.47)–(3.48) using the following relations:

Table 7.3 Numerical parameters of the active seat suspension model

System component	Parameter	Value	Unit
Pneumatic muscles	Nominal muscle diameter (d_{m0})	10×10^{-3}	m
	Nominal angle of the fibre (β_{m0})	25°	deg
	Nominal length of the muscle (l_{m0})	250×10^{-3}	m
	First parameter shaping force characteristics of the muscle (μ_m)	130	–
	Second parameter shaping force characteristics of the muscle (η_m)	35×10^{-3}	–
	Third parameter shaping force characteristics of the muscle (ν_m)	5	N
Proportional pressure control valves	Static gains of the proportional pressure control valves (k_{m1}, k_{m2})	1×10^5	Pa/V
	Time constants of the valves (t_{m1}, t_{m2})	0.01	s
	Minimum value of the control signal (u_{min})	0	V
	Maximum value of the control signal (u_{max})	10	V

$$p_{m1} = \frac{\mu_m \left(1 - \exp\left(\frac{l_{m0}+q_{1i}-q_{si}}{\eta_m l_{m0}}\right)\right) - \nu_m - F_{ax1}}{\frac{\pi}{4} d_{m1}^2 \left(1 - \frac{2}{\tan^2 \beta_{m1}}\right)} + p_0 \tag{7.27}$$

$$p_{m2} = \frac{\mu_m \left(1 - \exp\left(\frac{l_{m0}-(q_{1i}-q_{si})}{\eta_m l_{m0}}\right)\right) - \nu_m - F_{ax2}}{\frac{\pi}{4} d_{m2}^2 \left(1 - \frac{2}{\tan^2 \beta_{m2}}\right)} + p_0 \tag{7.28}$$

where F_{ax1} and F_{ax2} are the desired active force of particular muscles, d_{m1}, d_{m2} and β_{m1}, β_{m2} are the variable muscle diameters and angles of the fibre that are defined by Eq. (3.45), l_{m0} is the nominal length defined for a specific type of the pneumatic muscle, and μ_m, η_m and ν_m are the parameters shaping its force characteristics.

In Figs. 7.14 and 7.15, the graphical representation of the muscle inverse models is shown. It should be noted that the desired forces can be generated only within a specific workspace of the pneumatic muscle. If a large contraction of the pneumatic muscle is reached, then even a high pressure inside the muscle cannot provide the required active force.

In the following book, it is assumed that the controlled pressures p_{m1} and p_{m2} are proportional to the input signals u_{m1} and u_{m2} using the pressure control valves. In order to eliminate the actuating times t_{m1} and t_{m2} of control valves, the voltage input signals are subsequently formed by using the predictors PD as follows:

Fig. 7.17 Experimental set-up for the system verification of the active horizontal seat suspension

$$u_{m1} = \frac{1}{k_{m1}}(t_{m1}\dot{p}_{m1} + p_{m1}), \qquad u_{m2} = \frac{1}{k_{m2}}(t_{m2}\dot{p}_{m2} + p_{m2}) \tag{7.29}$$

where k_{m1} and k_{m2} are the static gains of the proportional pressure control valves, u_{m1} and u_{m2} are the predicted control signals. Such signals are finally restricted within the operating voltage range that is related to a specific type of pneumatic valves as:

$$u_{c1} = \begin{cases} u_{min} & \text{for} \quad u_{m1} < u_{min} \\ u_{m1} & \text{for} \quad u_{min} \le u_{m1} < u_{max} \\ u_{max} & \text{for} \quad u_{m1} \ge u_{max} \end{cases} \tag{7.30}$$

$$u_{c2} = \begin{cases} u_{min} & \text{for} \quad u_{m2} < u_{min} \\ u_{m2} & \text{for} \quad u_{min} \le u_{m2} < u_{max} \\ u_{max} & \text{for} \quad u_{m2} \ge u_{max} \end{cases} \tag{7.31}$$

where u_{min} and u_{max} are the minimum and maximum values of the signal that controls the proportional pressure control valves (typically 0–10 V).

7.7.3 Active System Verification

The unknown model parameters of the pneumatic muscles are determined adopting force–deflection measurements of the active system. A slow triangular motion is generated within total stroke of the seat suspension at a frequency of 0.1 Hz and an

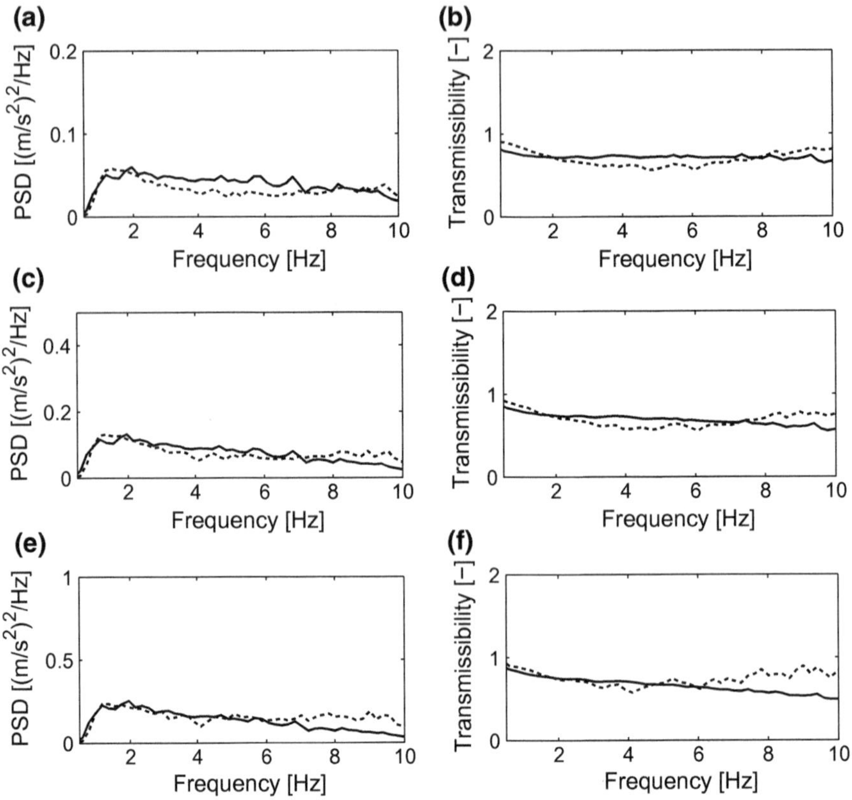

Fig. 7.18 Simulated (——) and measured (----) power spectral densities and transmissibility functions of the active horizontal seat suspension for different spectral classes: WN1x (**a**, **b**), WN2x (**c**, **d**), WN3x (**e**, **f**), mass load 30 kg

amplitude of 15 mm. Such a movement provides quasi-static conditions of the test, so the air conversion inside the pneumatic muscles can be treated as the isothermal process. The tests are carried out at constant pressure in both pneumatic muscles, i.e. $2 \cdot 10^5$, $4 \cdot 10^5$ and $9 \cdot 10^5$ Pa. The measured force in displacement domain is fitted to the simulation results (Fig. 7.16) by appropriate selection of the model parameters μ_m, η_m, ν_m. Numerical parameters of the active seat suspension model are shown in Table 7.3.

A verification of the system dynamics is performed by using the experimental set-up presented in Fig. 7.17. The active horizontal suspension is investigated for excitation signals similar to the white, band-limited noise in the range of frequency 1–10 Hz. Three different excitation intensities are used during the tests, i.e. WN1x ($1.02\ \mathrm{m/s^2}$), WN2x ($1.36\ \mathrm{m/s^2}$) and WN3x ($1.91\ \mathrm{m/s^2}$).

The measured signals are: the acceleration of vibration platform, the acceleration of isolated body, the relative displacement of suspension system and the absolute

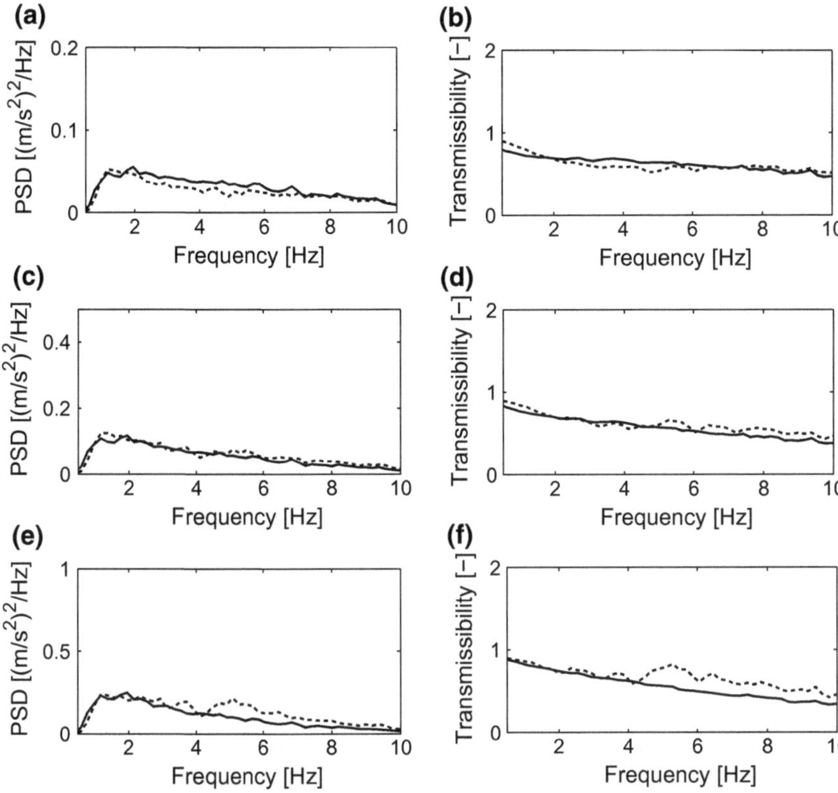

Fig. 7.19 Simulated (——) and measured (----) power spectral densities and transmissibility functions of the active horizontal seat suspension for different spectral classes: WN1x (**a**, **b**), WN2x (**c**, **d**), WN3x (**e**, **f**), mass load 55 kg

displacement of vibration platform. Based on these signals, the power spectral densities and transmissibility functions of the horizontal active suspension are evaluated. The corresponding results in the case of different loading masses (30, 55 and 80 kg) are shown in Figs. 7.18, 7.19 and 7.20.

The results of numerical simulations conducted on the horizontal seat suspension are close to the results measured experimentally. Based on a qualitative assessment of the obtained research results, the operation of an active system is found to be very effective in the whole frequency range, i.e. at frequencies from 1 to 10 Hz. The transmissibility functions are lower than 1 over the full range, which proves a correct functioning of the vibration control system.

In Table 7.4, simulated and measured transmissibility factors, suspension travels and relative errors of the active horizontal seat suspension are presented. As shown in this table, the maximum error does not exceed 9% for each of the excitation signals and at different mass loading. It proves a correct modelling of the suspension system which uses the pneumatic muscles in order to control the desired active force.

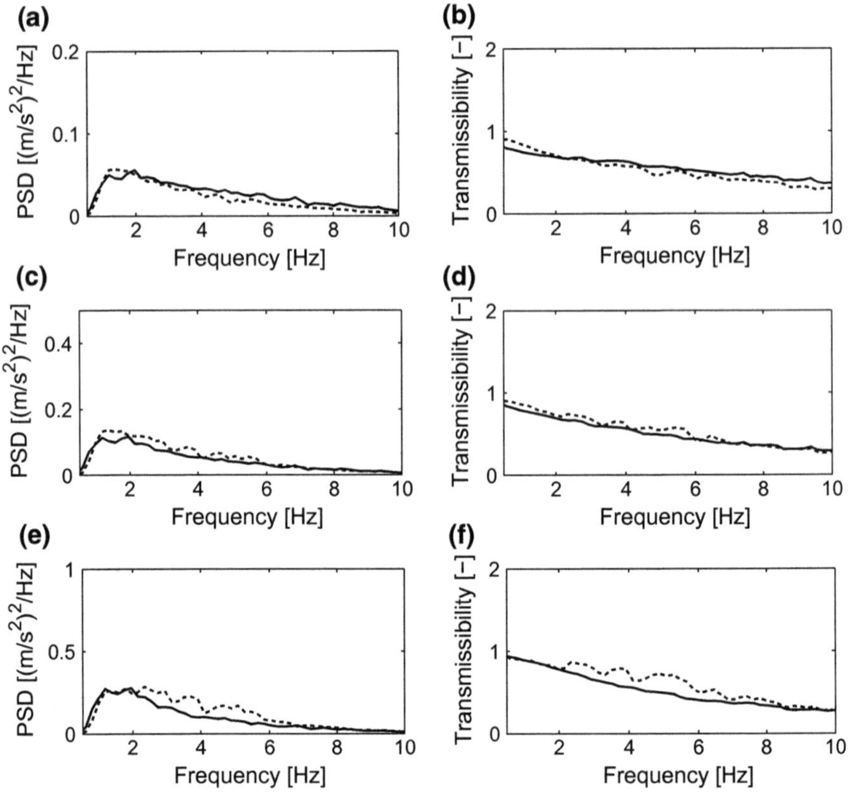

Fig. 7.20 Simulated (——) and measured (----) power spectral densities and transmissibility functions of the active horizontal seat suspension for different spectral classes: WN1x (**a**, **b**), WN2x (**c**, **d**), WN3x (**e**, **f**), mass load 80 kg

Table 7.4 Simulated and measured transmissibility factors, suspension travels and relative errors of the active horizontal seat suspension for excitation signals: WN1x, WN2x and WN3x

Suspended mass	Input vibration	Simulation		Measurement		Relative error	
		TFE_x factor	s_{tx} (mm)	TFE_x factor	s_{tx} (mm)	δ_{TFE_x} (%)	$\delta_{s_{tx}}$ (%)
30 kg	WN1x	0.728	18.3	0.714	17.4	1.9	5.1
	WN2x	0.739	23.4	0.712	22.6	3.7	3.5
	WN3x	0.737	27.1	0.744	26.2	1.0	3.4
55 kg	WN1x	0.681	21.3	0.665	20.6	2.4	3.3
	WN3x	0.675	26.2	0.683	25.9	1.1	1.1
	WN4x	0.707	32.1	0.741	29.6	4.5	8.4
80 kg	WN1x	0.668	23.0	0.670	22.0	0.2	4.5
	WN2x	0.662	30.4	0.699	27.9	5.2	8.9
	WN3x	0.726	32.6	0.779	30.9	6.8	5.5

7.8 Conclusions

In this chapter, an original control system structure is proposed for the purpose of improving the vibro-isolation properties of vibration reduction systems. The presented design methodology is based on the inverse model of control element and the primary controller that calculates the desired active force. Using the configurable control algorithm developed in this chapter, it is possible to achieve a significant reduction of vibrations transmitted into the isolated body or a substantial limiting of the suspension travel. The examples of control system synthesis are discussed especially for the semi-active horizontal seat suspension with MR damper and the active system with pneumatic muscles.

References

1. Konieczny, J., Kowal, J., Raczka, W., & Siebielak, M. (2013). Bench tests of slow and full active suspensions in terms of energy consumption. *Journal of low Frequency Noise Vibration and Active Control*, *32*(1–2), 81–98.
2. Zhang, Y., Zhang, X., Zhan, M., Guo, K., Zhao, F., & Liu, Z. (2015). Study on a novel hydraulic pumping regenerative suspension for vehicles. *Journal of the Franklin Institute*, *352*(2), 485–499.
3. Snamina, J., Kowal, J., & Orkisz, P. (2013). Active suspension based on low dynamic stiffness. *Acta Physica Polonica A*, *123*(6), 1118–1122.
4. Maciejewski, I., Meyer, L., & Krzyyski, T. (2010). The vibration damping effectiveness of an active seat suspension system and its robustness to varying mass loading. *Journal of Sound and Vibration*, *329*, 3898–3914.
5. Su, X., Yang, X., Shi, P., & Wu, L. (2014). Fuzzy control of nonlinear electromagnetic suspension systems. *Mechatronics*, *24*(4), 328–335.
6. Engel, Z., & Kowal, J. (1995). *Control of vibroacoustic processes (in Polish)*. Cracow: AGH University of Science and Technology Press.
7. Kovaleva, A. (2010). *Optimal control of mechanical oscillations*. Berlin, Heidelberg: Springer.
8. Astolfi, A., & Marconi, L. (2008). *Analysis and design of nonlinear control systems*. Berlin, Heidelberg: Springer.
9. Poussot-Vassal, C., Sename, O., Dugard, L., Gaspar, P., Szabo, Z., & Bokor, J. (2006). *Multi-objective qLPV Hinf/H2 control of a half vehicle*, 10th Mini Conference on vehicle system dynamics, identification and anomalies, VSDIA, Budapest, Hungary.
10. Poussot-Vassal, C., Sename, O., Dugard, L., Gaspar, P., Szabo, Z., & Bokor, J. (2008). A new semi-active suspension control strategy through LPV technique. *Control Engineering Practice*, *16*, 1519–1534.
11. Liu, Y., Matsuhisa, H., & Utsuno, H. (2008). Semi-active vibration isolation system with variable stiffness and damping control. *Journal of Sound and Vibration*, *313*, 16–28.
12. Yildirim, S. (2004). Vibration control of suspension systems using a proposed neural network. *Journal of Sound and Vibration*, *277*, 1059–1069.
13. Yoshimura, T., Nakaminami, K., Kurimoto, M., & Hino, J. (1999). Active suspension of passenger cars using linear and fuzzy-logic controls. *Control Engineering Practice*, *7*, 41–47.
14. Gu, D., Petkov, P., & Konstantinov, M. (2005). *Robust control design with MATLAB*. Berlin: Springer.
15. Maslanka, M., & Sapiski, B. (2007). Experimental study of vibration control of a cable with an attached MR damper. *Journal of Theoritical and Applied Mechanics*, *45*(4), 893–917.

16. Maciejewski, I. (2012). Control system design of active seat suspensions. *Journal of Sound and Vibration*, *331*, 1291–1309.
17. Maciejewski, I., Krzyzynski, T., & Meyer, L. (2014). Control system synthesis of seat suspensions used for protection of working machine operators. *Vehicle System Dynamics*, *52*(11), 1355–1371.
18. Choi, S. B., Lee, H. K., & Hang, E. G. (2001). Field test results of a semi-active ER suspension system associated with skyhook controller. *Mechatronics*, *11*, 345–353.

Chapter 8
A Method of Shaping the Vibro-isolation Properties of Semi-active and Active Systems

8.1 Nomenclature

$\mathbf{A}_i$	state (system) matrix (i = x, y, z)
$\mathbf{B}_{si}$, $\mathbf{B}_{ai}$	input matrices (i = x, y, z)
$\mathbf{C}_{1i}$, $\mathbf{C}_{2i}$	output matrices (i = x, y, z)
$\mathbf{D}_{1si}$, $\mathbf{D}_{1ai}$, $\mathbf{D}_{2si}$, $\mathbf{D}_{2ai}$	feedthrough (feedforward) matrices (i = x, y, z)
F_{ai}	applied force (i = x, y, z), N
$\mathbf{K}_i$	output feedback gain vector (i = x, y, z)
$\mathbf{q}_i$	displacement vector of the bio-mechanical model of human body (i = x, y, z)
q_{1i}	displacement of the isolated body for selected direction of the vibration exposure (i = x, y, z), m
q_{si}	displacement of the input vibration (i = x, y, z), m
$\mathbf{w}_{si}$	vector of the external disturbances (i = x, y, z)
$\mathbf{x}_d$	vector of the selected decision variables
$\mathbf{x}_i$	state vector (i = x, y, z)
$\mathbf{y}_i$	measurement vector (i = x, y, z)
$\mathbf{z}_i$	output vector (i = x, y, z)

8.2 Introduction

Passive vibration isolators are typically constructed using the inertial and visco-elastic elements. Although the dissipation of a substantial part of the vibration energy is provided at relatively high frequencies, the low-frequency vibration is typically intensified due to the resonance phenomena. As a result, it is difficult to achieve an efficient vibration isolation in a wide range of frequencies by means of the passive

T. Krzyzynski et al., *Modelling and Control Design of Vibration Reduction Systems*, https://doi.org/10.1007/978-3-030-03047-6_8

systems [1]. However, the increased vibration damping effectiveness can be obtained by applying semi-active or active systems [2].

In the following chapter, an original methodology of the control design is presented for the purpose of improving the suspension dynamics and thus effectively reducing the vibrations transmitted to the human body. An original system structure supports finding the control algorithm that yields a desired closed-loop response in view of different requirements for vibro-isolating properties of the modern vibration reduction systems.

8.3 Formulation of the Overall Method

The block diagram of the overall method for shaping vibro-isolation properties of semi-active and active systems is presented in Fig. 8.1. If the vibration exposure in typical working machines is simulated by means of random input signals, then the vibro-isolation properties have to be evaluated using simulation model of the vibration reduction system. Spectral characteristics of the input vibrations are specified in such a way that the generated excitation signals are representative for different types of working machines (see Chap. 2). Output signals of the simulation model should be used for determination and evaluation of risks from exposure to whole-body vibration (see Chap. 4). The desired vibro-isolation properties of semi-active or active system can be achieved by an appropriate selection of the controller settings that are calculated with the use of multi-criteria optimisation (see Chap. 5).

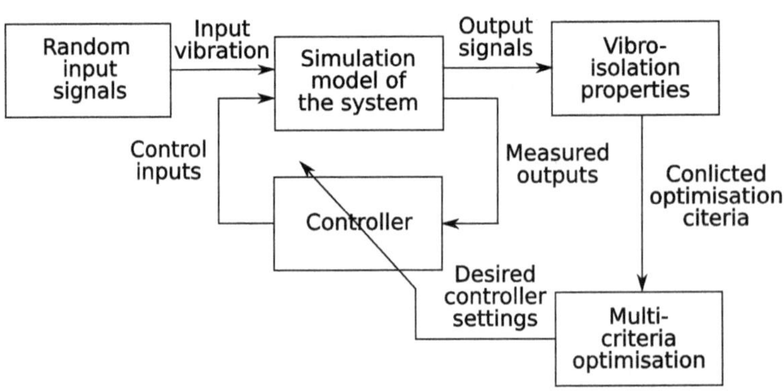

Fig. 8.1 Block diagram of the overall method for shaping vibro-isolation properties of semi-active and active systems

8.4 Simulation Model of the Semi-active and Active Systems

Simulation model of the semi-active or active vibration isolator should allow to predict the system performance under various input vibrations. If a model is a precise representation of the system dynamics, then an analysis of the control strategy can be successfully conducted by using numerical simulation. The system modelling equations and control algorithm have to be combined into one model in order to simulate the closed-loop system behaviour for different values of the controller settings. The

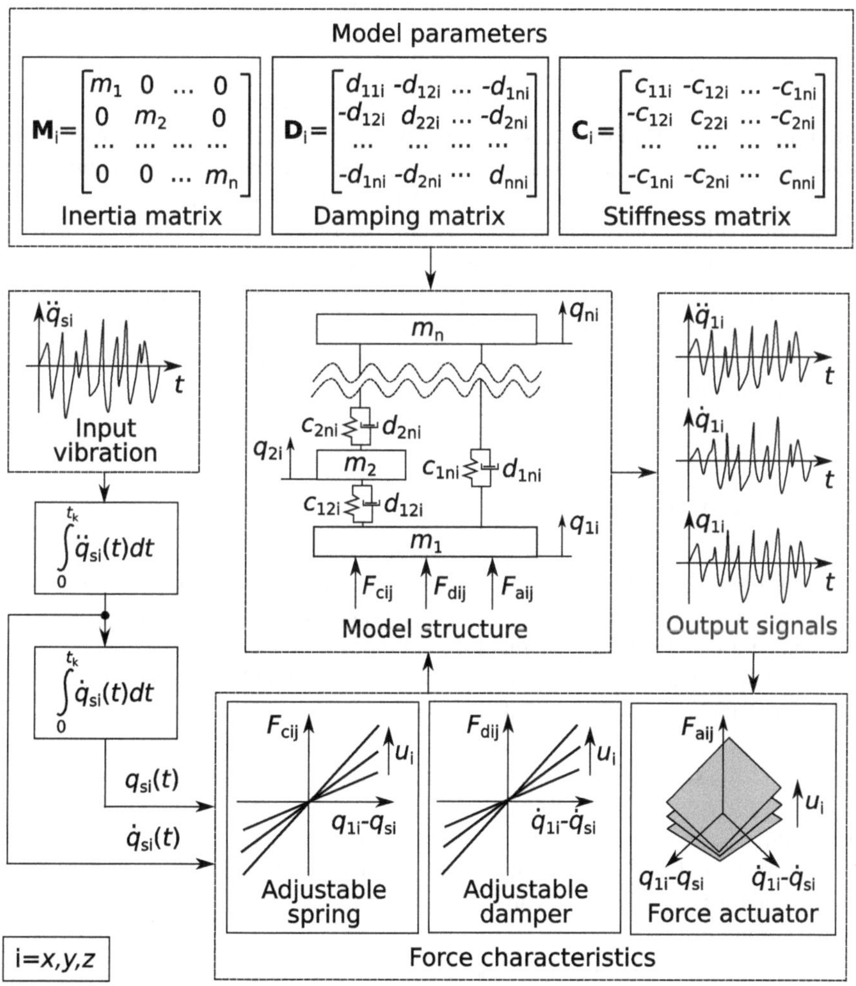

Fig. 8.2 Block diagram for numerical simulation of the dynamic behaviour of semi-active or active vibration reduction systems

block diagram representing a general simulation model of the semi-active or active vibration reduction systems is presented in Fig. 8.2.

Since human response to vibration shall be investigated for different spectral classes of the excitation signals, therefore the input vibration block (Fig. 8.2) is used to generate the random acceleration $\ddot{q}_{si}(t)$ for different directions of the vibration exposure ($i = x, y, z$). Then, a double integration with respect to time t is applied in order to obtain the velocity $\dot{q}_{si}(t)$ and displacement $q_{si}(t)$ as input signals to the model. The applied force of adjustable spring F_{cij}, adjustable damper F_{dij} or force actuator F_{aij} is employed to compensate harmful vibrations affecting the human body. Their non-linear characteristics are described as functions of the input signal u_i and the system relative displacement $q_{1i} - q_{si}$ and/or relative velocity $\dot{q}_{1i} - \dot{q}_{si}$. The force characteristics must be evaluated for the components used in a specific type of the semi-active or active suspension system (see Sect. 3.4).

If a model structure representing bio-dynamic response of the human body to single axis ($i = x, y, z$) is selected from the recent literature [3–5], then a set of differential equations of motion should be developed in the matrix form (Eq. 3.1). An appropriate selection of the model parameters specified as the inertia $\mathbf{M}_i$, damping $\mathbf{D}_i$ and stiffness $\mathbf{C}_i$ matrices (Eqs. 3.7–3.11) determines the displacements $q_{1i}, q_{2i}, \ldots, q_{ni}$ of the lumped mass body which typically consists of several interconnected masses by means of linear springs and dampers (see Sect. 3.7). The output acceleration $\ddot{q}_{1i}(t)$ is required in order to measure the human exposure to mechanical vibration and shock; however, the output velocity $\dot{q}_{1i}(t)$ and displacement $q_{1i}(t)$ are utilised for controlling vibrations of the suspension system with the use of adjustable spring, adjustable damper or force actuator.

8.5 Control System

The controller is determined on the basis of generalised system model that is shown in Fig. 8.3. The presented model provides the versatility of created control algorithm that can be used successfully in the different types of suspension systems.

The state variables represent the movement of elements contained in the bio-mechanical model of a human body as follows:

$$\mathbf{x}_i(t) := [q_{1i}, \dot{q}_{1i}, q_{2i}, \dot{q}_{2i}, \ldots, q_{ni}, \dot{q}_{ni}]^T, \qquad i = x, y, z \tag{8.1}$$

where $q_{1i}, q_{2i}, \ldots, q_{ni}$ and $\dot{q}_{1i}, \dot{q}_{2i}, \ldots, \dot{q}_{ni}$ are the displacements and velocities of the human body model (Fig. 8.3). The displacement q_{si} and velocity $\dot{q}_{si}$ are defined as the external disturbances, e.g. cab vibrations of the machine which is moving over uneven ground:

$$\mathbf{w}_{si}(t) := [q_{si}, \dot{q}_{si}]^T, \qquad i = x, y, z \tag{8.2}$$

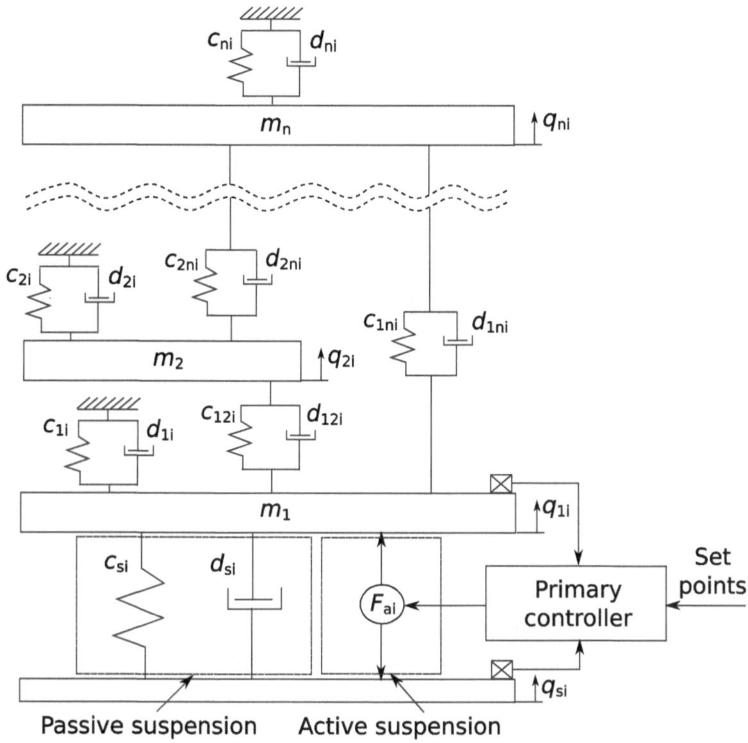

Fig. 8.3 Generalised model of the hybrid vibration reduction system for one direction of the vibration transmission (i $= x, y, z$)

Assuming the applied force F_{ai} to be an output from the primary controller, the state-space equation of hybrid vibration reduction system can be obtained by using the linear fractional transformation (LFT) technique [6]:

$$\dot{\mathbf{x}}_{\mathrm{i}}(t) = \mathbf{A}_{\mathrm{i}}\mathbf{x}_{\mathrm{i}}(t) + \mathbf{B}_{\mathrm{si}}\mathbf{w}_{\mathrm{si}}(t) + \mathbf{B}_{\mathrm{ai}}F_{\mathrm{ai}}(t), \qquad \mathrm{i} = x, y, z \tag{8.3}$$

The state (system) matrix is expressed as follows:

$$\mathbf{A}_{\mathrm{i}} = \begin{bmatrix} 0 & 1 & 0 & 0 & \cdots & 0 & 0 \\ -\frac{c_{11\mathrm{i}}}{m_1} & -\frac{d_{11\mathrm{i}}}{m_1} & \frac{c_{12\mathrm{i}}}{m_1} & \frac{d_{12\mathrm{i}}}{m_1} & \cdots & \frac{c_{1\mathrm{ni}}}{m_1} & \frac{d_{1\mathrm{ni}}}{m_1} \\ 0 & 0 & 0 & 1 & \cdots & 0 & 0 \\ \frac{c_{12\mathrm{i}}}{m_2} & \frac{d_{12\mathrm{i}}}{m_2} & -\frac{c_{22\mathrm{i}}}{m_2} & -\frac{d_{22\mathrm{i}}}{m_2} & \cdots & \frac{c_{2\mathrm{ni}}}{m_2} & \frac{d_{2\mathrm{ni}}}{m_2} \\ \cdots & \cdots & \cdots & \cdots & \cdots & \cdots & \cdots \\ 0 & 0 & 0 & 0 & \cdots & 0 & 1 \\ \frac{c_{1\mathrm{ni}}}{m_\mathrm{n}} & \frac{d_{1\mathrm{ni}}}{m_\mathrm{n}} & \frac{c_{2\mathrm{ni}}}{m_\mathrm{n}} & \frac{d_{2\mathrm{ni}}}{m_\mathrm{n}} & \cdots & -\frac{c_{\mathrm{nni}}}{m_\mathrm{n}} & -\frac{d_{\mathrm{nni}}}{m_\mathrm{n}} \end{bmatrix} \tag{8.4}$$

The input matrices are given using the following relations:

$$\mathbf{B}_{\mathrm{si}} = \begin{bmatrix} 0 & 0 \\ \frac{c_{\mathrm{si}}}{m_1} & \frac{d_{\mathrm{si}}}{m_1} \\ 0 & 0 \\ 0 & 0 \\ \cdots & \cdots \\ 0 & 0 \\ 0 & 0 \end{bmatrix} \qquad \mathbf{B}_{\mathrm{ai}} = \begin{bmatrix} 0 \\ \frac{1}{m_1} \\ 0 \\ 0 \\ \cdots \\ 0 \\ 0 \end{bmatrix} \tag{8.5}$$

The selected damping $d_{11\mathrm{i}}, d_{22\mathrm{i}}, \ldots, d_{\mathrm{nni}}$ and stiffness $c_{11\mathrm{i}}, c_{22\mathrm{i}}, \ldots, c_{\mathrm{nni}}$ coefficients should be calculated as follows:

$$\begin{aligned} d_{11\mathrm{i}} &= d_{1\mathrm{i}} + d_{12\mathrm{i}} + \cdots + d_{1\mathrm{ni}} + d_{\mathrm{si}} & c_{11\mathrm{i}} &= c_{1\mathrm{i}} + c_{12\mathrm{i}} + \cdots + c_{1\mathrm{ni}} + c_{\mathrm{si}} \\ d_{22\mathrm{i}} &= d_{2\mathrm{i}} + d_{12\mathrm{i}} + \cdots + d_{2\mathrm{ni}} & c_{22\mathrm{i}} &= c_{2\mathrm{i}} + c_{12\mathrm{i}} + \cdots + c_{2\mathrm{ni}} \\ &\cdots & &\cdots \\ d_{\mathrm{nni}} &= d_{\mathrm{ni}} + d_{1\mathrm{ni}} + \cdots + d_{(\mathrm{n}-1)\mathrm{ni}} & c_{\mathrm{nni}} &= c_{\mathrm{ni}} + c_{1\mathrm{ni}} + \cdots + c_{(\mathrm{n}-1)\mathrm{ni}} \end{aligned} \tag{8.6}$$

If the isolated body acceleration $\ddot{q}_{1\mathrm{i}}$ and the suspension travel $q_{1\mathrm{i}} - q_{\mathrm{si}}$ are defined as the output variables, then the output vector can be written as:

$$\mathbf{z}_{\mathrm{i}}(t) := [\ddot{q}_{1\mathrm{i}}, q_{1\mathrm{i}} - q_{\mathrm{si}}]^{\mathrm{T}}, \qquad \mathrm{i} = x, y, z \tag{8.7}$$

Therefore, the output equation reads:

$$\mathbf{z}_{\mathrm{i}}(t) = \mathbf{C}_{1\mathrm{i}}\mathbf{x}_{\mathrm{i}}(t) + \mathbf{D}_{1\mathrm{si}}\mathbf{w}_{\mathrm{si}}(t) + \mathbf{D}_{1\mathrm{ai}}F_{\mathrm{ai}}(t), \qquad \mathrm{i} = x, y, z \tag{8.8}$$

with the corresponding output matrix:

$$\mathbf{C}_{1\mathrm{i}} = \begin{bmatrix} -\frac{c_{11\mathrm{i}}}{m_1} & -\frac{d_{11\mathrm{i}}}{m_1} & \frac{c_{12\mathrm{i}}}{m_1} & \frac{d_{12\mathrm{i}}}{m_1} & \cdots & \frac{c_{1\mathrm{ni}}}{m_1} & \frac{d_{1\mathrm{ni}}}{m_1} \\ 1 & 0 & 0 & 0 & \cdots & 0 & 0 \end{bmatrix} \tag{8.9}$$

The feedforward matrices are defined in the following manner:

$$\mathbf{D}_{1\mathrm{si}} = \begin{bmatrix} \frac{c_{\mathrm{si}}}{m_1} & \frac{d_{\mathrm{si}}}{m_1} \\ -1 & 0 \end{bmatrix} \qquad \mathbf{D}_{1\mathrm{ai}} = \begin{bmatrix} \frac{1}{m_1} \\ 0 \end{bmatrix} \tag{8.10}$$

The suspension travel $q_{1\mathrm{i}} - q_{\mathrm{si}}$ and the absolute velocity $\dot{q}_{1\mathrm{i}}$ of isolated body should be measurable; therefore, the measurement vector can be expressed as:

$$\mathbf{y}_{\mathrm{i}}(t) := [q_{1\mathrm{i}} - q_{\mathrm{si}}, \dot{q}_{1\mathrm{i}}]^{\mathrm{T}}, \qquad \mathrm{i} = x, y, z \tag{8.11}$$

The corresponding measurement equation reads:

$$\mathbf{y}_i(t) = \mathbf{C}_{2i}\mathbf{x}_i(t) + \mathbf{D}_{2si}\mathbf{w}_{si}(t) + \mathbf{D}_{2ai}F_{ai}(t), \qquad i = x, y, z \tag{8.12}$$

with the following matrices:

$$\mathbf{C}_{2i} = \begin{bmatrix} 1\,0\,0\,0 \cdots 0\,0 \\ 0\,1\,0\,0 \cdots 0\,0 \end{bmatrix} \tag{8.13}$$

$$\mathbf{D}_{2si} = \begin{bmatrix} -1 & 0 \\ 0 & 0 \end{bmatrix} \qquad \mathbf{D}_{2ai} = \begin{bmatrix} 0 \\ 0 \end{bmatrix} \tag{8.14}$$

A real-time control problem is formulated using the measurement vector $\mathbf{y}_i(t)$ and multiplying it by the output feedback gain vector $\mathbf{K}_i = [k_{1i}, k_{2i}]$ in the following order:

$$F_{ai}(t) = \mathbf{K}_i\mathbf{y}_i(t), \qquad i = x, y, z \tag{8.15}$$

where k_{1i} is the proportionality factor of the relative displacement feedback loop, and k_{2i} is the proportionality factor of the absolute velocity feedback loop.

8.6 Example: Control System Design of the Horizontal Suspension with Seated Human Body

8.6.1 Physical and Mathematical Model of the System

In Fig. 8.4, physical and mathematical model of the horizontal seat suspension with seated human body is shown. An elementary 3-DOF model is employed to represent the system bio-dynamic response under harmful vibrations [4]. In order to evaluate the desired force that should be generated in the system actively, the linear spring c_{sx} and damper d_{sx} are utilised to describe essential characteristics of the horizontal seat suspension.

The linear suspension system can be modelled using a state-space approach for the purpose of the primary controller synthesis. The following state variables are used to reproduce the movement of a human body in the longitudinal x-direction:

$$\mathbf{x}_x(t) := [q_{1x}, \dot{q}_{1x}, q_{2x}, \dot{q}_{2x}, q_{3x}, \dot{q}_{3x}]^T \tag{8.16}$$

where q_{1x}, q_{2x}, q_{3x} and $\dot{q}_{1x}, \dot{q}_{2s}, \dot{q}_{3x}$ are the displacements and velocities of the bio-mechanical model (Fig. 8.4). The external disturbances are defined as the displacement q_{sx} and velocity $\dot{q}_{sx}$ of input vibration:

$$\mathbf{w}_{sx}(t) := [q_{sx}, \dot{q}_{sx}]^T \tag{8.17}$$

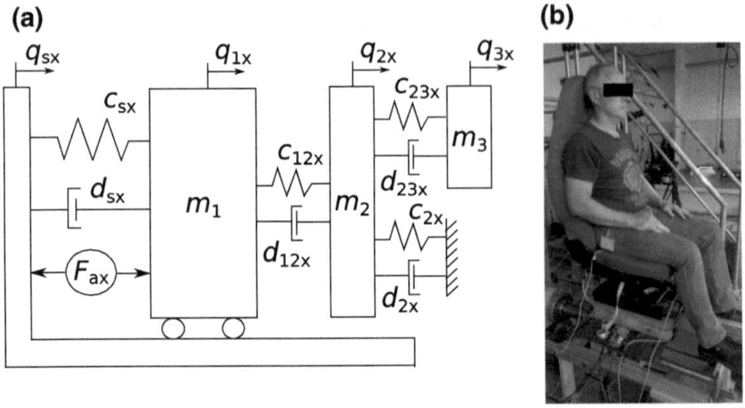

Fig. 8.4 Physical model of the active suspension with seated human body (**a**) and experimental test stand (**b**)

The state-space equation of the active horizontal suspension is obtained using the applied force F_{ax} as an output from the primary controller:

$$\dot{\mathbf{x}}_{\mathrm{x}}(t) = \mathbf{A}_{\mathrm{x}}\mathbf{x}_{\mathrm{x}}(t) + \mathbf{B}_{\mathrm{sx}}\mathbf{w}_{\mathrm{sx}}(t) + \mathbf{B}_{\mathrm{ax}}F_{\mathrm{ax}}(t) \tag{8.18}$$

The state (system) matrix is formulated in the following form:

$$\mathbf{A}_{\mathrm{x}} = \begin{bmatrix} 0 & 1 & 0 & 0 & 0 & 0 \\ -\frac{c_{11\mathrm{x}}}{m_1} & -\frac{d_{11\mathrm{x}}}{m_1} & \frac{c_{12\mathrm{x}}}{m_1} & \frac{d_{12\mathrm{x}}}{m_1} & 0 & 0 \\ 0 & 0 & 0 & 1 & 0 & 0 \\ \frac{c_{12\mathrm{x}}}{m_2} & \frac{d_{12\mathrm{x}}}{m_2} & -\frac{c_{22\mathrm{x}}}{m_2} & -\frac{d_{22\mathrm{x}}}{m_2} & \frac{c_{2\mathrm{ni}}}{m_2} & \frac{d_{2\mathrm{ni}}}{m_2} \\ 0 & 0 & 0 & 0 & 0 & 1 \\ 0 & 0 & \frac{c_{23\mathrm{x}}}{m_3} & \frac{d_{23\mathrm{x}}}{m_3} & -\frac{c_{33\mathrm{x}}}{m_3} & -\frac{d_{33\mathrm{x}}}{m_3} \end{bmatrix} \tag{8.19}$$

The following input matrices are given for the seat suspension system:

$$\mathbf{B}_{\mathrm{sx}} = \begin{bmatrix} 0 & 0 \\ \frac{c_{\mathrm{sx}}}{m_1} & \frac{d_{\mathrm{sx}}}{m_1} \\ 0 & 0 \\ 0 & 0 \\ 0 & 0 \\ 0 & 0 \end{bmatrix} \qquad \mathbf{B}_{\mathrm{ax}} = \begin{bmatrix} 0 \\ \frac{1}{m_1} \\ 0 \\ 0 \\ 0 \\ 0 \end{bmatrix} \tag{8.20}$$

The selected damping d_{11x}, d_{22x}, d_{33x} and stiffness c_{11x}, c_{22x}, c_{33x} coefficients should be calculated using the following relations:

$$\begin{aligned} d_{11x} &= d_{12x} + d_{sx} & c_{11x} &= c_{12x} + c_{sx} \\ d_{22x} &= d_{2x} + d_{12x} + d_{23x} & c_{22x} &= c_{2x} + c_{12x} + c_{23x} \\ d_{33i} &= d_{23x} & c_{33x} &= c_{23x} \end{aligned} \tag{8.21}$$

The output vector contains the isolated body acceleration $\ddot{q}_{1x}$ and the suspension travel $q_{1x} - q_{sx}$ in order to assess the system performance:

$$\mathbf{z}_x(t) := [\ddot{q}_{1x}, q_{1x} - q_{sx}]^T \tag{8.22}$$

Using such a system configuration, the output equation can be written as:

$$\mathbf{z}_x(t) = \mathbf{C}_{1x}\mathbf{x}_x(t) + \mathbf{D}_{1sx}\mathbf{w}_{sx}(t) + \mathbf{D}_{1ax}F_{ax}(t) \tag{8.23}$$

where the corresponding output matrix is described as follows:

$$\mathbf{C}_{1x} = \begin{bmatrix} -\frac{c_{11x}}{m_1} & -\frac{d_{11x}}{m_1} & \frac{c_{12x}}{m_1} & \frac{d_{12x}}{m_1} & 0 & 0 \\ 1 & 0 & 0 & 0 & 0 & 0 \end{bmatrix} \tag{8.24}$$

In succession, the feedforward matrices are given as:

$$\mathbf{D}_{1sx} = \begin{bmatrix} \frac{c_{sx}}{m_1} & \frac{d_{sx}}{m_1} \\ -1 & 0 \end{bmatrix} \qquad \mathbf{D}_{1ax} = \begin{bmatrix} \frac{1}{m_1} \\ 0 \end{bmatrix} \tag{8.25}$$

The system configuration presented in this book requires a measuring of the suspension travel $q_{1x} - q_{sx}$ and the absolute velocity $\dot{q}_{1x}$ of isolated body. Therefore, the measurement vector can be expressed as:

$$\mathbf{y}_x(t) := [q_{1x} - q_{sx}, \dot{q}_{1x}]^T \tag{8.26}$$

The corresponding measurement equation reads:

$$\mathbf{y}_x(t) = \mathbf{C}_{2x}\mathbf{x}_x(t) + \mathbf{D}_{2sx}\mathbf{w}_{sx}(t) + \mathbf{D}_{2ax}F_{ax}(t) \tag{8.27}$$

with the following matrices:

$$\mathbf{C}_{2x} = \begin{bmatrix} 1 & 0 & 0 & 0 & 0 & 0 \\ 0 & 1 & 0 & 0 & 0 & 0 \end{bmatrix} \tag{8.28}$$

$$\mathbf{D}_{2sx} = \begin{bmatrix} -1 & 0 \\ 0 & 0 \end{bmatrix} \qquad \mathbf{D}_{2ax} = \begin{bmatrix} 0 \\ 0 \end{bmatrix} \tag{8.29}$$

The desired active force is finally determined by using the following relation:

$$F_{ax}(t) = \mathbf{K}_x \mathbf{y}_x(t) \tag{8.30}$$

where $\mathbf{K}_x = [k_{1x}, k_{2x}]$ is the output feedback gain vector to be designed.

8.6.2 Optimisation of the Controller Settings

The proportionality factor $x_{d1} := k_{1x}$ of the relative displacement feedback loop and the proportionality factor $x_{d2} := k_{2x}$ of the absolute velocity feedback loop are

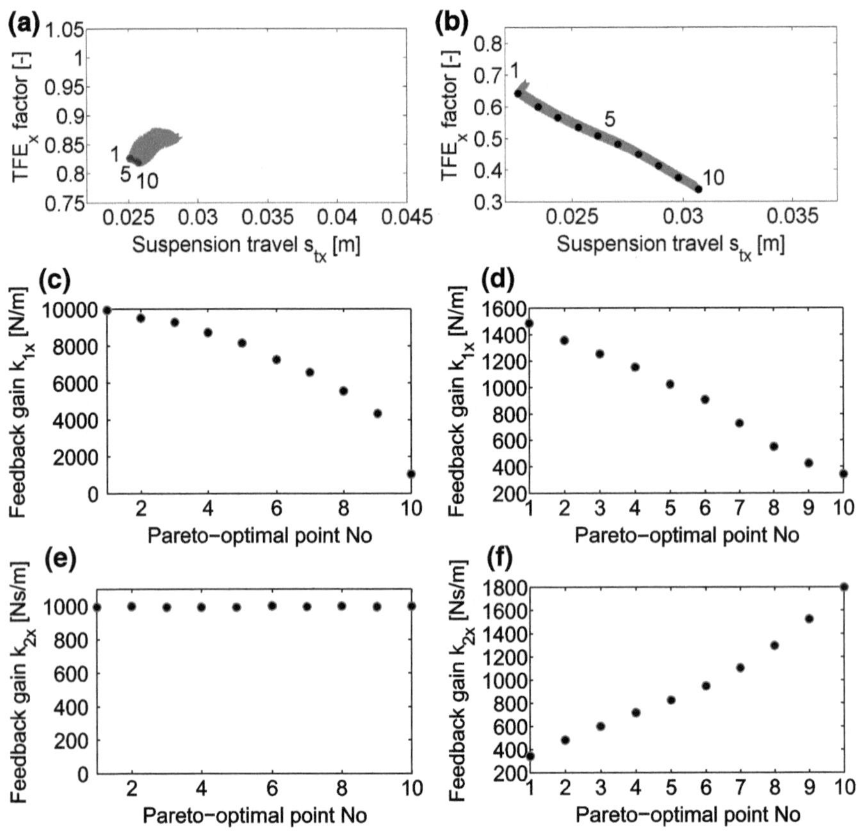

Fig. 8.5 Pareto-optimal point distribution for the semi-active (**a**) and active (**b**) seat suspensions, corresponding primary controller settings obtained for the semi-active (**c, e**) and active (**d, f**) systems, input vibration WN1x

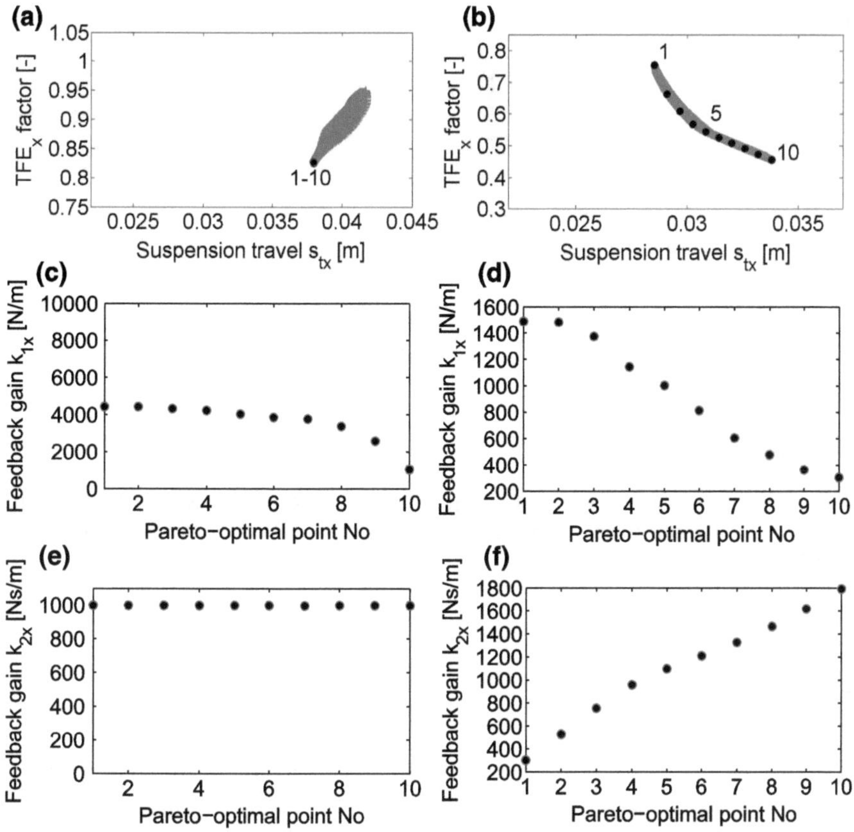

Fig. 8.6 Pareto-optimal point distribution for the semi-active (**a**) and active (**b**) seat suspensions, corresponding primary controller settings obtained for the semi-active (**c**, **e**) and active (**d**, **f**) systems, input vibration WN2x

selected as meaningful decision variables that have the greatest influence on system dynamics. The corresponding vector of the chosen decision variables reads:

$$\mathbf{x}_\mathrm{d} = [x_{\mathrm{d}1}, x_{\mathrm{d}2}]^\mathrm{T} \tag{8.31}$$

For the purpose of controlling the human vibration, the transmissibility factor $\mathrm{TFE_x}$ is utilised as a principal optimisation criterion. The less important criterion (suspension travel s_{tx}) is considered as the non-linear inequality constraint during optimisation process. The optimisation procedure is realised using randomly starting points within the variability ranges that are defined for the particular controller settings.

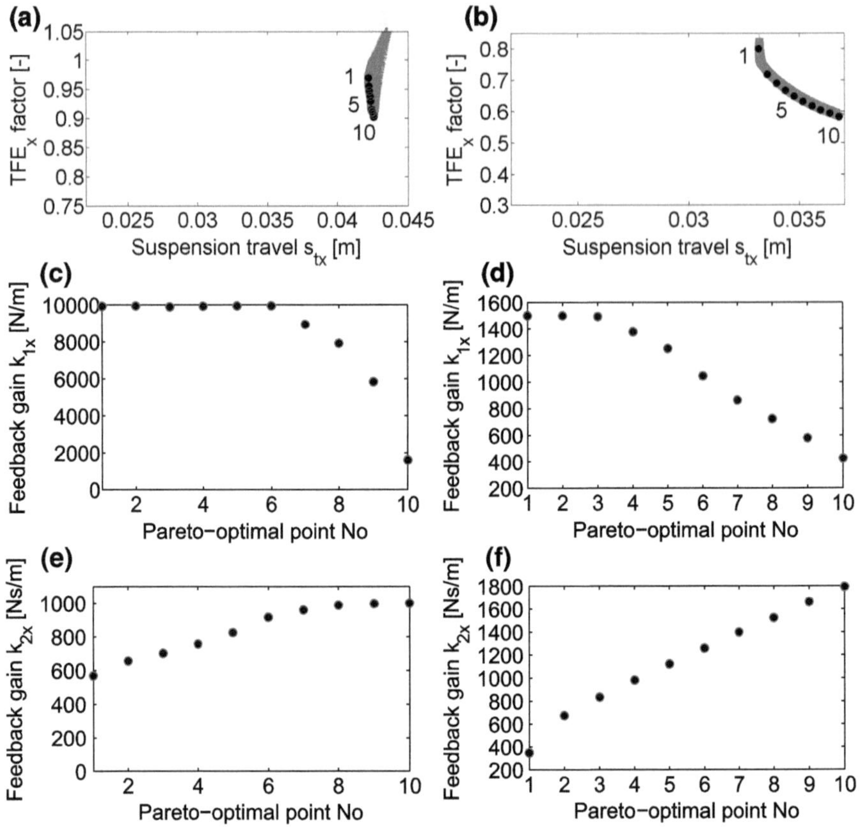

Fig. 8.7 Pareto-optimal point distribution for the semi-active (**a**) and active (**b**) seat suspensions, corresponding primary controller settings obtained for the semi-active (**c**, **e**) and active (**d**, **f**) systems, input vibration WN3x

The following ranges are taken in the case of a semi-active system with the MR damper:

- proportionality factor of the relative displacement feedback loop
  $k_{1x} = 1000 - 10000\ \mathrm{N/m}$,
- proportionality factor of the absolute velocity feedback loop
  $k_{2x} = 100 - 1000\ \mathrm{Ns/m}$.

In succession, the controller settings for an active system with the pneumatic muscles are defined as:

- proportionality factor of the relative displacement feedback loop
  $k_{1x} = 300 - 1500\ \mathrm{N/m}$,
- proportionality factor of the absolute velocity feedback loop
  $k_{2x} = 300 - 1800\ \mathrm{Ns/m}$.

Ten Pareto-optimal solutions are found for each of the excitation signals used, i.e. the signal similar to white, band-limited noise in the frequency range 0.5–10 Hz at different excitation intensities: 1.02 m/s^2 (WN1x), 1.36 m/s^2 (WN2x), 1.91 m/s^2 (WN3x). The optimisation results are shown in Figs. 8.5, 8.6 and 8.7. The obtained Pareto-optimal solutions are expressed by different sets of the decision variables that correspond to the primary controller settings.

These optimisation results (Figs. 8.5, 8.6 and 8.7) clearly show that a change of the primary controller settings: k_{1x} and k_{2x} within their variability ranges involves modifications of the vibration transmissibility behaviour. An increase of the first controller

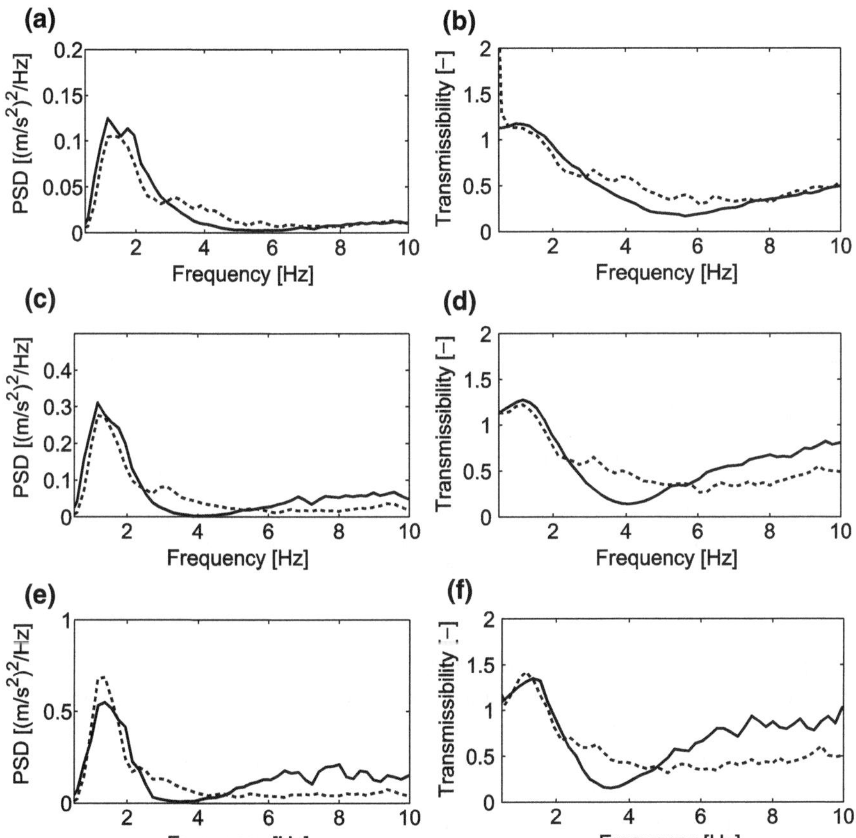

Fig. 8.8 Simulated (—) and measured (----) power spectral densities of the semi-active suspension for different spectral classes: WN1x (**a**), WN2x (**c**), WN3x (**e**) and transmissibility functions for different spectral classes: WN1x (**b**), WN2x (**d**), WN3x (**f**), human body mass 90 kg

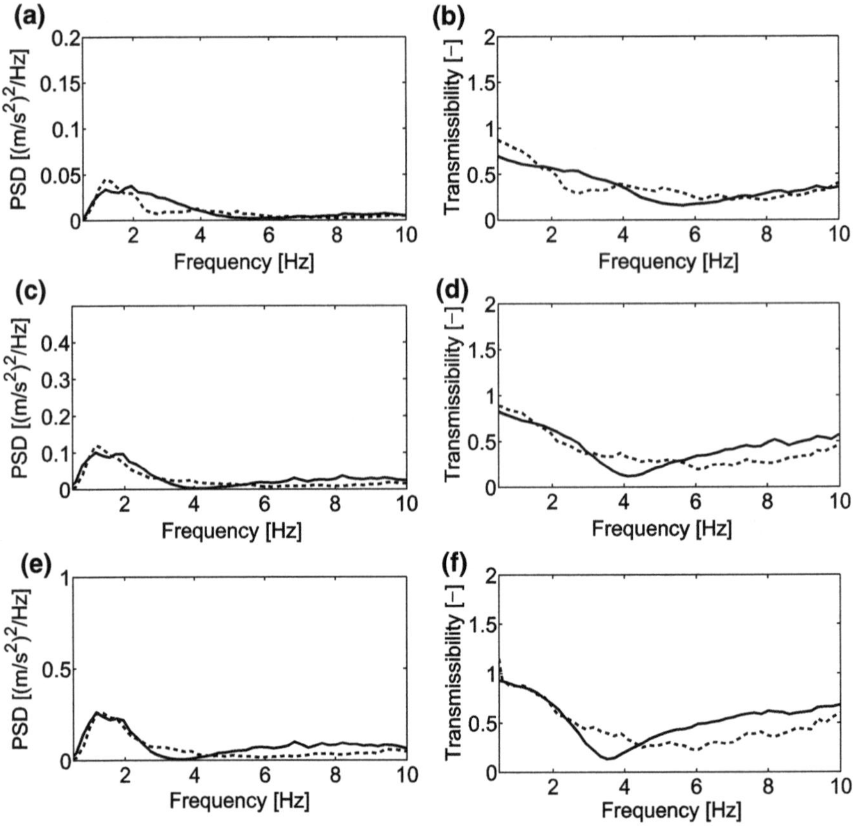

Fig. 8.9 Simulated (—) and measured (----) power spectral densities of the active suspension for different spectral classes: WN1x (**a**), WN2x (**c**), WN3x (**e**) and transmissibility functions for different spectral classes: WN1x (**b**), WN2x (**d**), WN3x (**f**), human body mass 90 kg

setting k_{1x} causes less system activity in reducing harmful vibrations transmitted to the human body; however, a limitation of the suspension travel s_{tx} can be achieved. Although an increase of the second controller setting k_{2x} contributes lowering of the transmissibility factor TFE_x, a relative displacement of the suspension system also increases. The presented results show that an appropriate selection of the primary controller settings (defined by Pareto-optimal solutions) allows adjusting the vibro-isolation properties of semi-active and active seats in view of the opposite requirements for modern suspension systems.

8.6.3 Laboratory Investigation of the Semi-active and Active Seat Vibration Control

The laboratory investigations of the semi-active and active seat vibration control are performed by using the NI cRIO-9074 integrated system that combines a real-time processor for embedded machine control and monitoring applications. Both the primary and secondary controllers are designed using a graphical programming language LabVIEW that allows to create sophisticated controllers and deploy the control system to real-time hardware.

The actual suspension deflection is evaluated by a cable extension position sensor, which measures a relative displacement between the isolated body and the seat base. The velocity of the suspended body is obtained by integrating and filtering the acceleration signal which is measured using a linear accelerometer [7]. The MR damper developed at the Koszalin University of Technology is used to control the seat vibration (semi-active system). Besides that, two high-speed proportional pressure valves (Rexroth type MNR) are employed to control the air-pressure inside the pneumatic muscles (active system). The muscles can generate the active force that is required to reduce harmful vibrations experienced by the human body.

The power spectral densities and transmissibility functions are evaluated for different spectral classes of the excitation signals, i.e. WN1x, WN2x and WN3x. The corresponding simulation and measurement results are presented in Figs. 8.8 and 8.9.

As shown in Fig. 8.8, the semi-active system unsatisfactorily decreases the vibration amplitudes at low frequencies (below 2 Hz) of the excitation signals that is noticeable especially for the high-intensity vibrations, i.e. WN2x and WN3x. The reason for this is insufficient range of the MR damper force at different currents flowing through the coil windings (0–4 A). The corresponding force characteristics in velocity domain have been presented in the previous chapter (Fig. 7.8).

The active suspension is very effective in reducing vibrations (Fig. 8.9) that adversely affect human health at work. Both simulated and measured transmissibility functions indicate low values (less than 1) in the frequency range 0.5–10 Hz reflecting the proper operation of the active vibration reduction system. Moreover, the spectral characteristics measured experimentally are compatible with the simulation results satisfactorily.

In Tables 8.1 and 8.2, simulated and measured transmissibility factors, suspension travels and relative errors of the semi-active and active horizontal seat suspension with seated human body are presented. As shown in this table, in the case of 90 kg person the maximum error does not exceed 10% for each of the excitation signals used during the tests.

Table 8.1 Simulated and measured transmissibility factors, suspension travels and relative errors of the semi-active horizontal suspension with seated human body obtained for excitation signals: WN1x, WN2x and WN3x, human body mass 90 kg

Input vibration	Simulation		Measurement		Relative error	
	TFE_x factor	s_{tx}, mm	TFE_x factor	s_{tx}, mm	δ_{TFE_x}, %	$\delta_{s_{tx}}$, %
WN1x	0.848	28.1	0.810	25.8	4.6	8.9
WN2x	0.861	35.3	0.810	38.5	6.2	8.4
WN3x	0.883	39.2	0.891	43.5	0.9	9.9

Table 8.2 Simulated and measured transmissibility factors, suspension travels and relative errors of the active horizontal suspension with seated human body obtained for excitation signals: WN1x, WN2x and WN3x, human body mass 90 kg

Input vibration	Simulation		Measurement		Relative error	
	TFE_x factor	s_{tx}, mm	TFE_x factor	s_{tx}, mm	δ_{TFE_x}, %	$\delta_{s_{tx}}$, %
WN1x	0.514	26.2	0.507	26.5	1.3	1.1
WN2x	0.573	31.2	0.551	30.0	3.9	4.0
WN3x	0.641	34.9	0.620	32.5	3.3	7.3

8.7 Selection of the Vibro-isolation Properties of Semi-active and Active Systems for Chosen Excitation Signals

The controller settings $x_{d1} := k_{1x}$ and $x_{d2} := k_{2x}$ are selected once more as the set of decision variables $\mathbf{x}_d = [x_{d1}, x_{d2}]^T$. In order to optimise the suspension system vibro-isolating properties, minimisation of the first criterion $TFE_x(\mathbf{x}_d)$ (frequency-weighted transmissibility factor) is carried out; however, the second criterion $s_{tx}(\mathbf{x}_d)$ (suspension travel) is again treated as a non-linear inequality constraint. Such defined task is solved by the optimisation algorithm that attempts to find a minimum of the objective function starting at the previously defined constraint value s_{txj}. The results of optimisation procedure for the selected input vibrations, ie. AL1x, AT1x, AT2x, AT3x, AT4x and RV1x, are shown in Figs. 8.10 and 8.11.

In Figs. 8.12 and 8.13, optimal controller settings of the semi-active suspension with MR damper are presented. In succession, optimal controller settings of the active suspension with pneumatic muscles are shown in Figs. 8.14 and 8.15. These controller settings are obtained for particular Pareto-optimal solutions that define different system configurations. An appropriate selection of the Pareto-optimal point

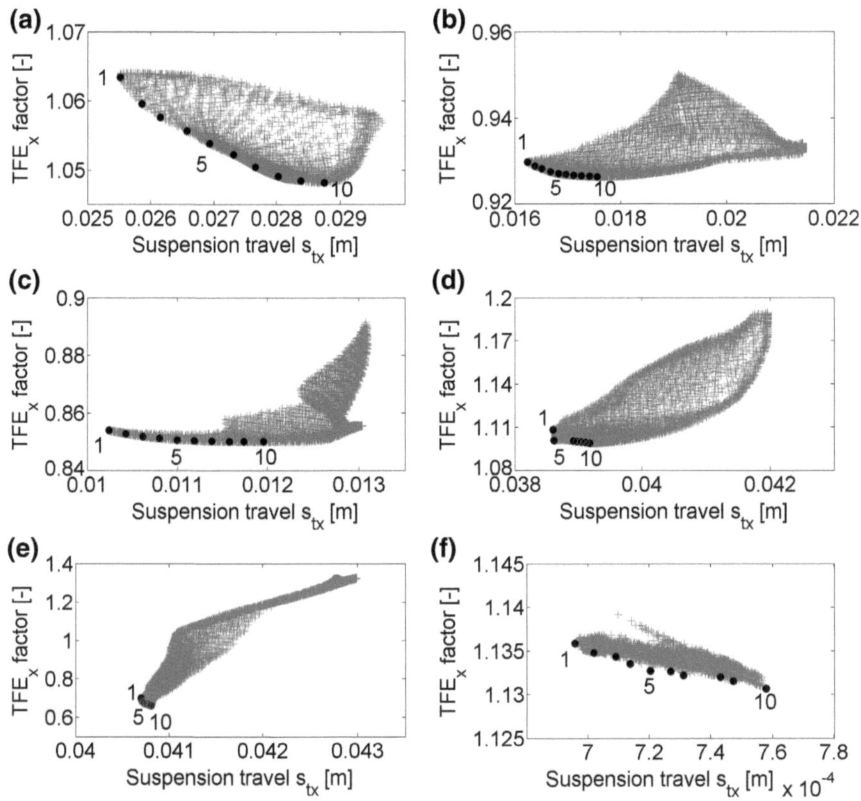

Fig. 8.10 Pareto-optimal point distribution (•) of semi-active system for the selected input vibration: AL1x (**a**), AT1x (**b**), AT2x (**c**), AT3x (**d**), AT4x (**e**), RV1x (**f**)

No. allows to change the system dynamics for the purpose of substantially limiting the suspension travel s_{tx} (point No. 1) or considerably reducing the vibration transmitted to a human body (point No. 10). The compromising semi-active and active system configurations are defined by the Pareto-optimal solutions from No. 2 up to No. 9.

Unlike the previously considered passive system, the controller settings for semi-active (Figs. 8.12, 8.13) and active (Figs. 8.14, 8.15) are directly available and they should be programmed in the control system of seat suspension. Values of these controller settings correspond to the various Pareto-optimal solutions that are calculated relative to the opposing vibro-isolation criteria.

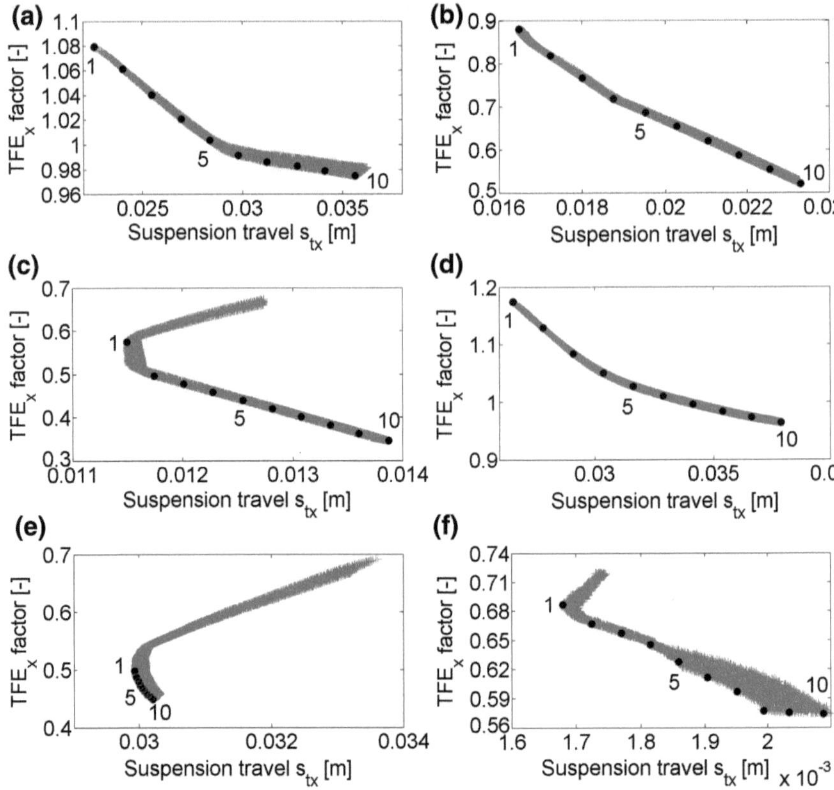

Fig. 8.11 Pareto-optimal point distribution (•) of active system for the selected input vibration: AL1x (**a**), AT1x (**b**), AT2x (**c**), AT3x (**d**), AT4x (**e**), RV1x (**f**)

Transmissibility functions of the semi-active and active seat suspension for different spectral classes, i.e. AL1x, AT1x, AT2x, AT3x, AT4x and RV1x, are presented in Fig. 8.16. As follows from these results, the effectiveness of active seat suspension is higher compared to the semi-active system. The highest efficiency of active system is observed in the frequency range 0.5–4 Hz. In this range, the pneumatic muscles considerably reduce whole-body vibration that is transmitted to the seat in horizontal direction. The semi-active system equipped with MR damper is not powerful enough to absorb the harmful vibration energy released by moving the heavy machinery on uneven ground. The seat suspension dynamic behaviour of both systems, i.e. semi-active and active, is similar to each other in the frequency range higher than 4 Hz.

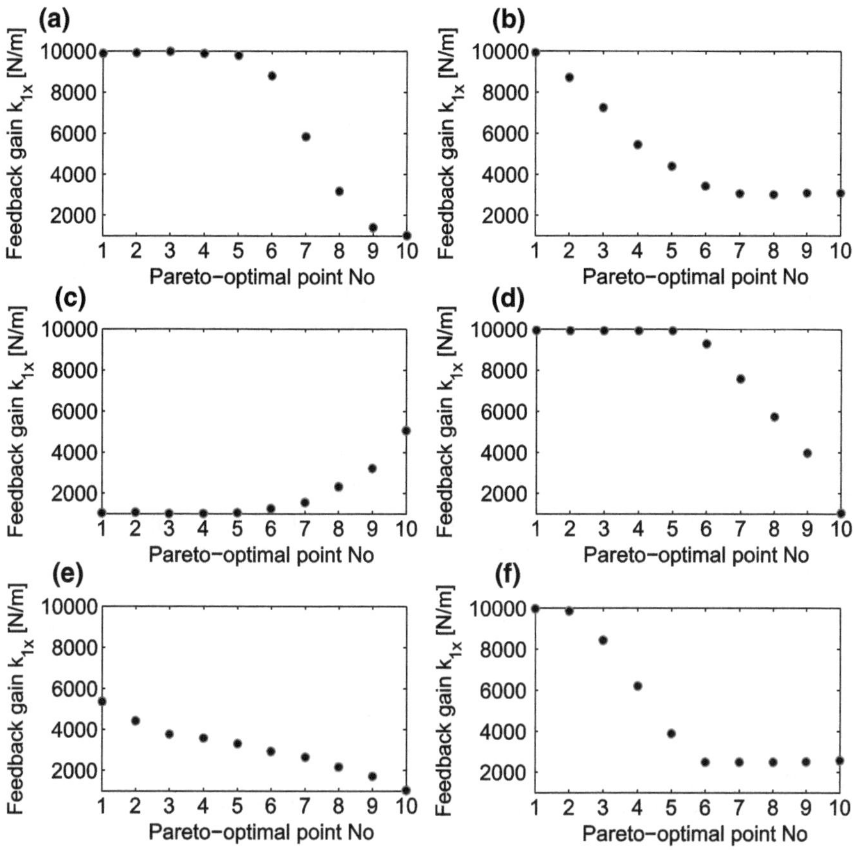

Fig. 8.12 Optimal controller setting k_{1x} of the semi-active suspension with MR damper obtained for the following input vibration: AL1x (**a**), AT1x (**b**), AT2x (**c**), AT3x (**d**), AT4x (**e**) and RV1x (**f**)

The transmissibility factors TFE_x measured for the active suspension at different input signals are much lower in comparison with the results measured for the semi-active system (Table 8.3). This proves the high efficiency of the pneumatic system equipped with two artificial muscles acting opposite to each other. However, in some cases the suspension stroke is required larger in order to limit vibrations that adversely affect the work of machine operators.

In Fig. 8.17, the root mean square accelerations are shown for different exposure to vibration inputs. The presented results are investigated under laboratory conditions over a measurement period of 300 s, and they are expressed as a function of the exposure duration. As shown in the presented figures, the daily exposure limit value (1.15 m/s^2) is not exceeded by using semi-active or active seat suspension systems.

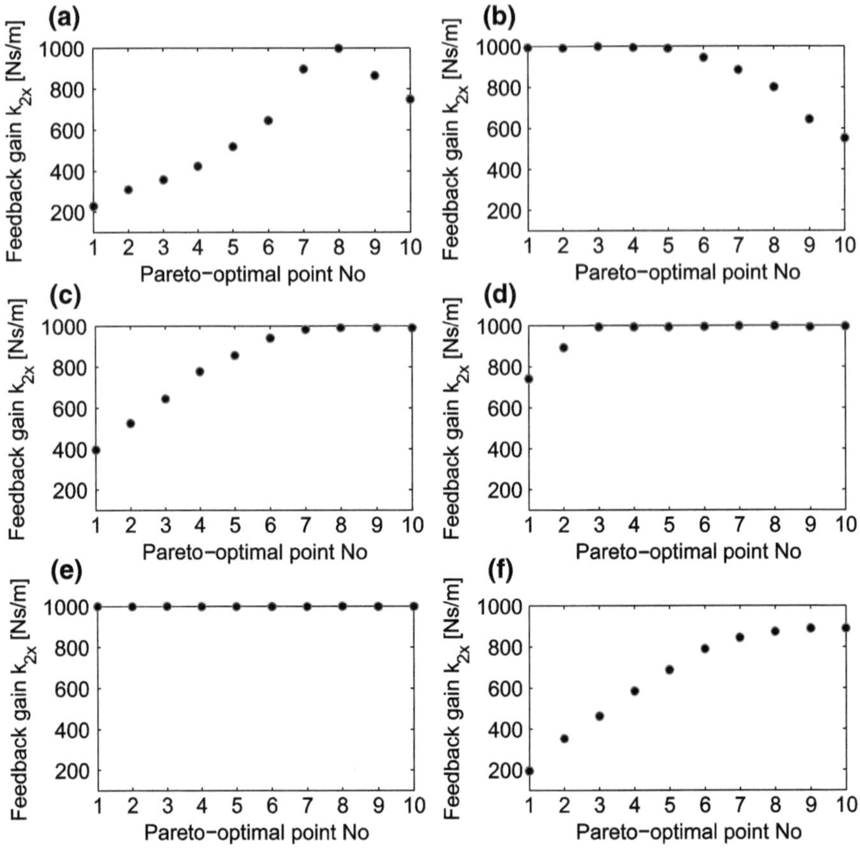

Fig. 8.13 Optimal controller setting k_{2x} of the semi-active suspension with MR damper obtained for the following input vibration: AL1x (**a**), AT1x (**b**), AT2x (**c**), AT3x (**d**), AT4x (**e**) and RV1x (**f**)

However, the daily exposure action value (0.5 m/s^2) is surpassed for the excitation signals AT3x and AT4x at relatively large vibration magnitudes. The vibration dose values (Fig. 8.18) are always below the daily exposure action value (9.1 $m/s^{1.75}$); therefore, the risk from exposure to vibration and shock seems to be low.

8.8 Conclusions

In this chapter, a method of shaping the vibro-isolation properties of semi-active and active systems is discussed. A generalised model of the vibration reduction system is developed by using a state-space approach. Then, the controller synthesis is

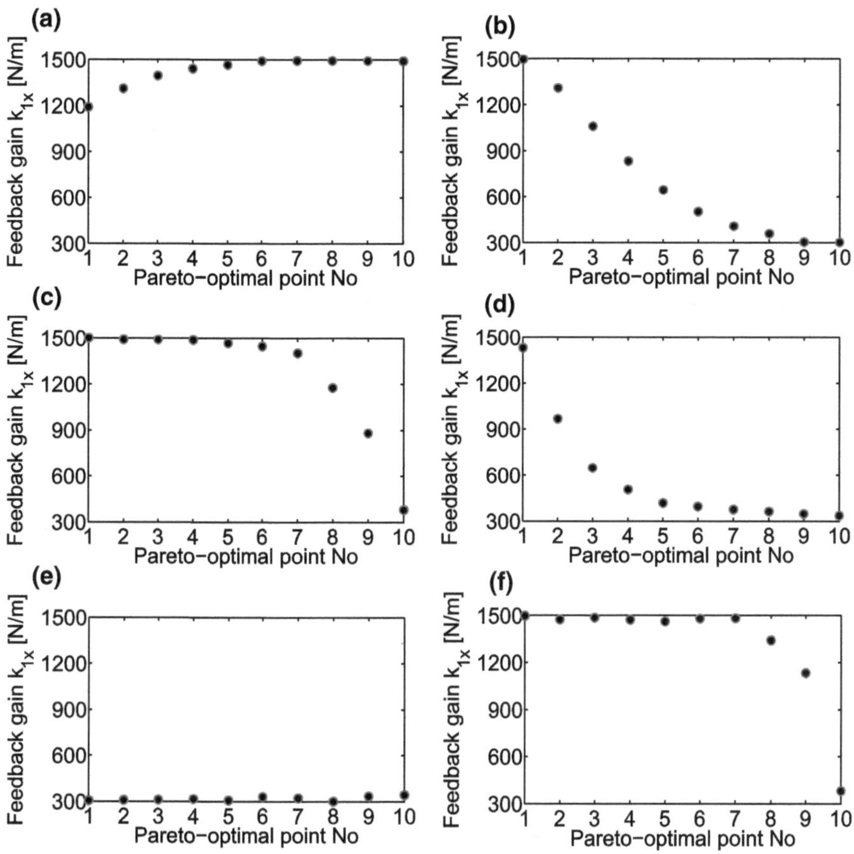

Fig. 8.14 Optimal controller setting k_{1x} of the active suspension with pneumatic muscles obtained for the following input vibration: AL1x (**a**), AT1x (**b**), AT2x (**c**), AT3x (**d**), AT4x (**e**) and RV1x (**f**)

performed by formulating the state feedback control problem. The output feedback gain vector is designed employing the multi-criteria optimisation that allows to restrict the conflicted vibro-isolation criteria. A correctness of the proposed method is tested experimentally using the exemplary vibration reduction systems, i.e. semi-active and active horizontal seat suspensions. Their vibro-isolation properties are selected for specific input vibrations that are representative for different working machines.

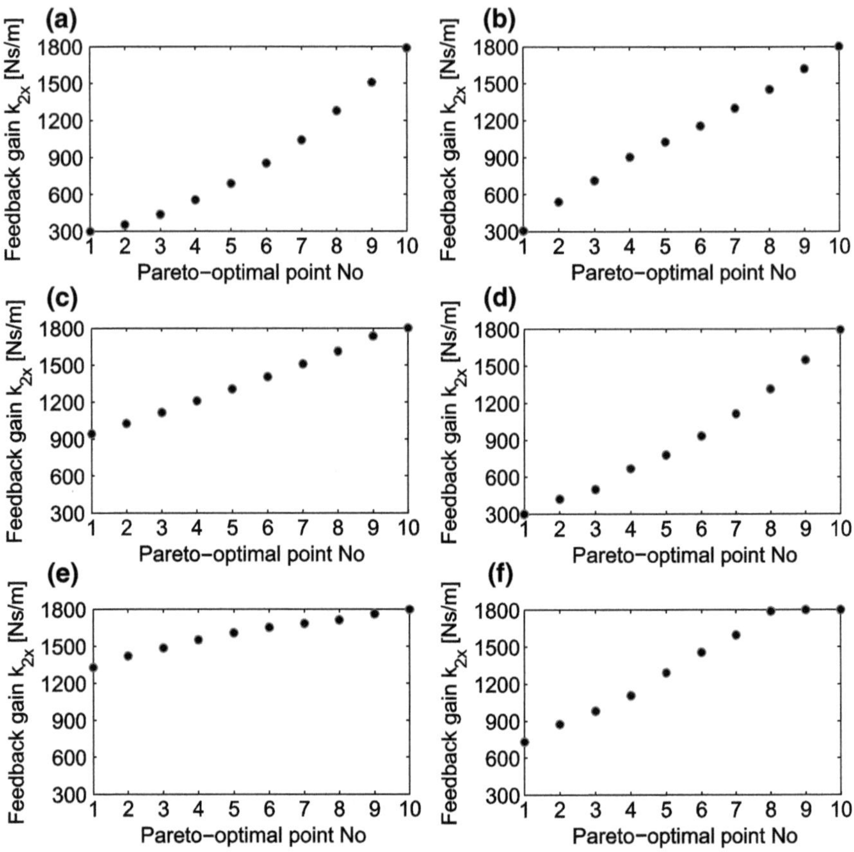

Fig. 8.15 Optimal controller setting k_{2x} of the active suspension with pneumatic muscles obtained for the following input vibration: AL1x (**a**), AT1x (**b**), AT2x (**c**), AT3x (**d**), AT4x (**e**) and RV1x (**f**)

Table 8.3 Numerical values of the transmissibility factors and suspension travels obtained for excitation signals: AL1x, AT1x, AT2x, AT3x, AT4x and RV1x, human body mass 90 kg

Input vibration	Semi-active		Active	
	TFE_x factor	s_{tx}, mm	TFE_x factor	s_{tx}, mm
AL1x	1.048	28.3	0.976	35.9
AT1x	0.922	17.2	0.521	23.3
AT2x	0.854	12.3	0.338	13.9
AT3x	1.104	39.0	0.961	37.6
AT4x	0.657	40.8	0.442	30.1
RV1x	1.129	0.7	0.564	2.1

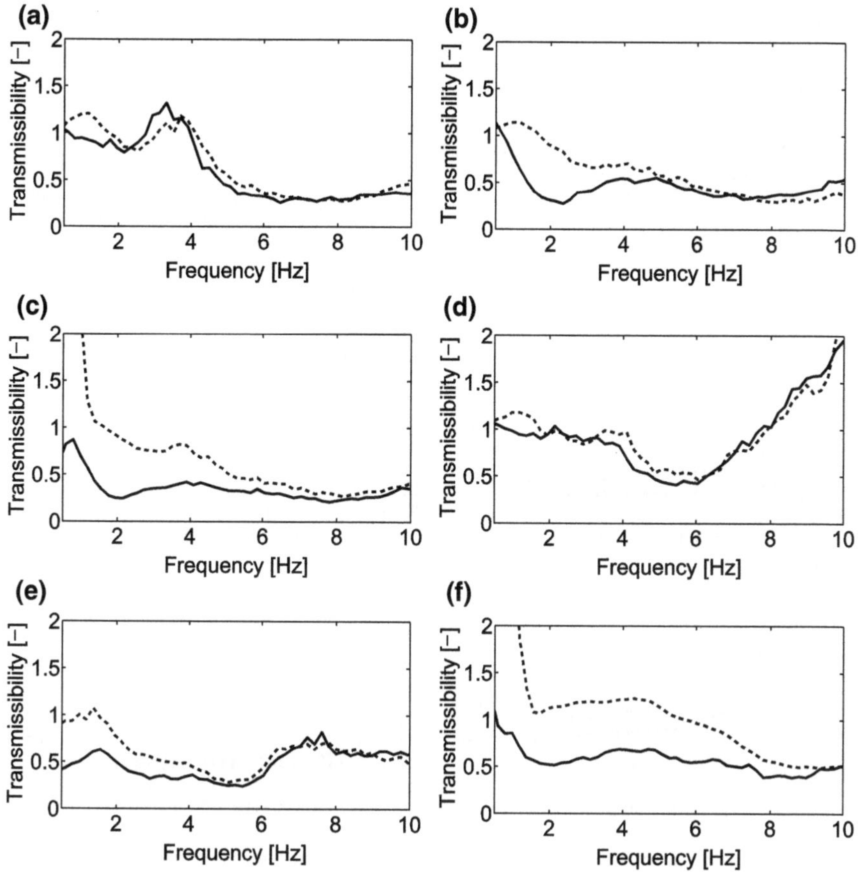

Fig. 8.16 Transmissibility functions of the semi-active (----) and active (——) horizontal seat suspension for different spectral classes: AL1x (**a**), AT1x (**b**), AT2x (**c**), AT3x (**d**), AT4x (**e**), RV1x (**f**), human body mass 90 kg

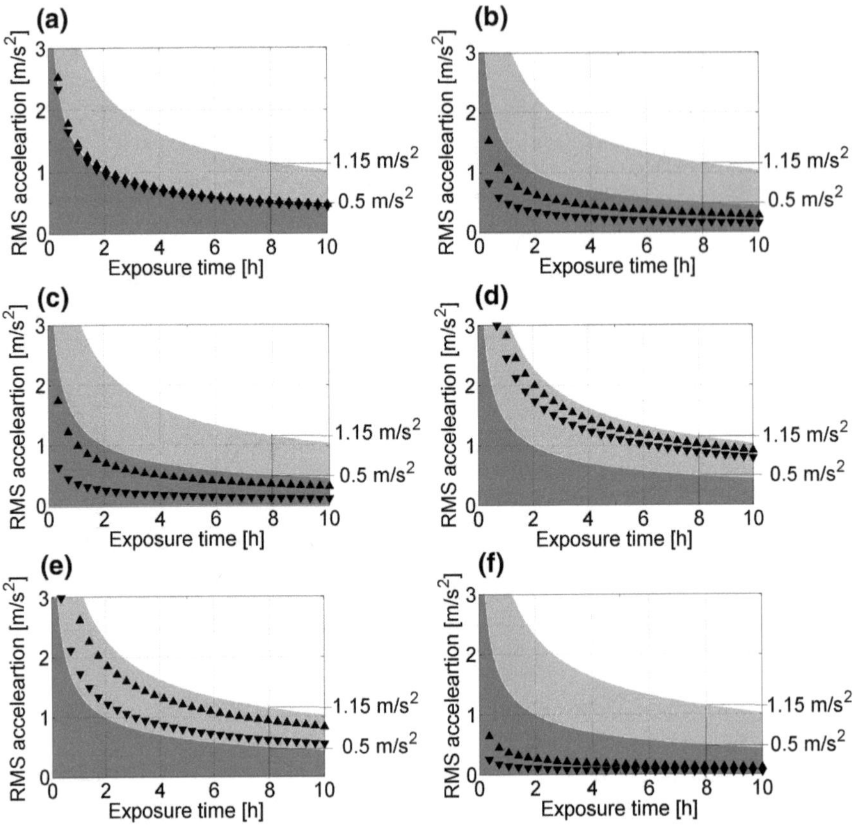

Fig. 8.17 Root mean square accelerations for different exposure to input vibrations: AL1x (**a**), AT1x (**b**), AT2x (**c**), AT3x (**d**), AT4x (**e**), RV1x (**f**) obtained using the semi-active (▲) and active (▼) horizontal seat suspension, human body mass 90 kg

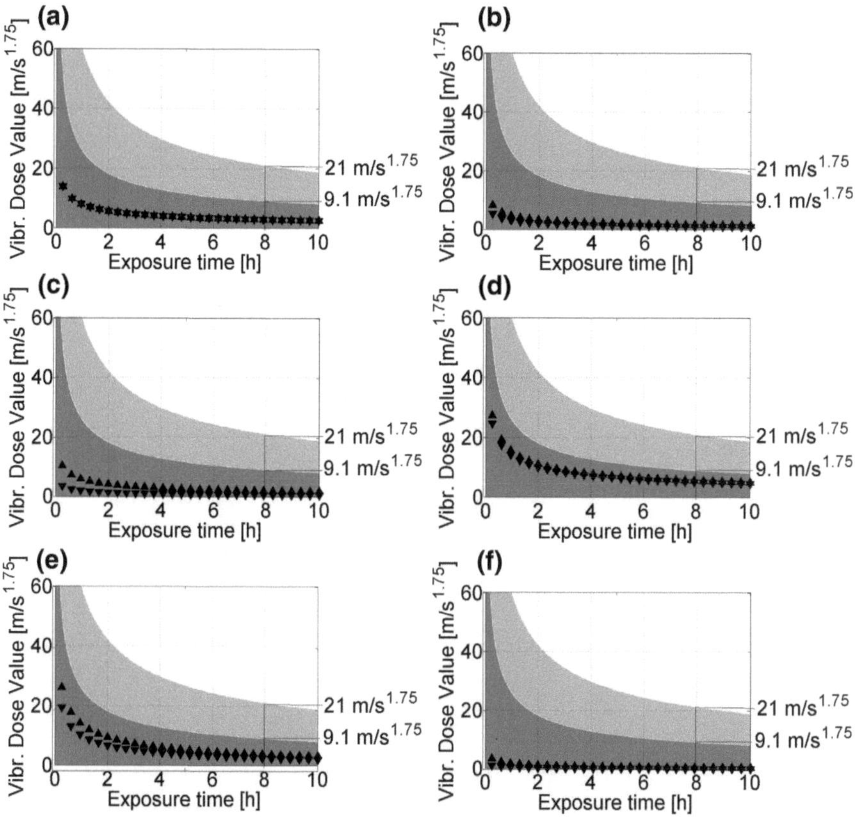

Fig. 8.18 Vibration dose values for different exposure to input vibrations: AL1x (**a**), AT1x (**b**), AT2x (**c**), AT3x (**d**), AT4x (**e**), RV1x (**f**) obtained using the semi-active (▲) and active (▼) horizontal seat suspension, human body mass 90 kg

References

1. Maciejewski, I., Meyer, L., & Krzyzynski, T. (2009). Modelling and multi-criteria optimisation of passive seat suspension vibro-isolating properties. *Journal of Sound and Vibration*, *324*, 520–538.
2. Maciejewski, I., Meyer, L., & Krzyzynski, T. (2010). The vibration damping effectiveness of an active seat suspension system and its robustness to varying mass loading. *Journal of Sound and Vibration*, *329*, 3898–3914.
3. Rutzel, S., Hinz, B., & Wolfel, H. B. (2006). Modal description–A better way of characterizing human vibration behavior. *Journal of Sound and Vibration*, *298*, 810–823.
4. Stein, G. J., Muka, P., Chmurny, R., Hinz, B., & Bluthner, R. (2007). Measurement and modelling of x-direction apparent mass of the seated human body–cushioned seat system. *Journal of Biomechanics*, *40*, 1493–1503.
5. Toward, M., & Griffin, J. (2011). The transmission of vertical vibration through seats: Influence of the characteristics of the human body. *Journal of Sound and Vibration*, *330*, 6526–6543.
6. Gu, D., Petkov, P., & Konstantinov, M. (2005). *Robust control design with MATLAB*. Berlin: Springer.

7. Lin, Y. C., & Khalil, H. K. (1992). Two-scale design of active suspension control using acceleration feedback. (pp. 884–889).

Chapter 9
Conclusions

9.1 Summary of the Most Important Results

The book presents the research results concerning the shaping vibration isolation properties of vibration reduction systems in order to reduce the harmful influence of vibrations on working machine operators during their work. The research methodology assumed the formulation of generalised procedures to enable their application in the process of designing of vibration isolation systems.

First of all, theoretical models of random signals are determined, whose spectral characteristics are representative of vibrations most often affecting machine operators. This formed the basis to define proper excitations of vibration reduction systems and thus made it possible to reproduce their real working conditions. Next, the general model of vibration reduction system and isolated object in the form of discrete system with many degrees of freedom is designed. The possibility of modification of non-linear characteristics is introduced by changing the parameters of models which describe the basic visco-elastic elements applied in the vibration isolator. Also, the conflicted assessment criteria of the effectiveness of operation of vibration reduction systems are devised. Criteria values are determined during computer-simulated vibration isolation process for a strictly defiled conditions of the excitation signal and the system load.

For such a formulated problem, the methodology of shaping vibration isolation properties of passive vibration reduction systems is developed. The methodology enables to determine non-linear characteristics of the basic visco-elastic elements used in the system. This aim is achieved by using the optimisation procedures, proposed by the authors, which have the ability to find the compromise solutions (Pareto-optimal) for the conflicted vibration isolation criteria.

In further part of the book, the methodology of shaping vibration isolation properties of technically advanced semi-active and active systems is discussed. The developed strategy helps to determine the structure of the control algorithm which is based on the inverse model of actuators and on the primary controller of a vibration

T. Krzyzynski et al., *Modelling and Control Design of Vibration Reduction Systems*, https://doi.org/10.1007/978-3-030-03047-6_9

reduction system. The optimisation procedure used enables to find Pareto-optimal solutions of control systems by proper selection of the primary controller settings.

Furthermore, the correctness of the proposed methodology is verified by performing experimental tests of real vibration reduction systems. The tests are executed with the use of the optimised characteristics of basic visco-elastic elements applied in the passive system and with the use of optimised controller settings in the case of semi-active and active system. For all the analysed vibration isolation systems: passive, semi-active and active, satisfactory agreement of the experimental and theoretical investigations is achieved, when actuating the seat suspension motion with random signals at different spectral characteristics.

The application of the developed methodology for the selection of the vibration isolation properties of suspension systems enabled to achieve effective vibration reduction at high frequencies of excitation in slight amplification of resonant amplitude. Additionally, lower vibration amplitudes of an isolated body are obtained, which considerably improve proper vibration isolation. The proposed research activities should help to make a choice between a desirable reduction of harmful mechanical vibrations and conflicted reduction of the suspension travel. Such an approach contributes to the improvement of comfort of a machine operator apart from having a beneficial influence on the contact of the machine operator with steering systems inside the cabin.

9.2 Final Conclusions

The aim of the book presented in the introduction is achieved by providing original methodology which enables to determine the best vibration isolation properties of suspension systems. The general conclusions regarding the developed models of vibration reduction systems and shaping their vibration isolation properties are presented in the following points.

9.2.1 Conclusions Regarding the Developed Models of Vibration Reduction Systems

The most important general conclusions regarding the developed models of vibration reduction systems are the following:

- the application of computer modelling techniques and numerical methods of optimisations enables to design modern vibration reduction systems with the use of non-linear models,

- the application of non-linear visco-elastic characteristics of vibration reduction systems contributes to the improvement of their vibration isolation properties in relation to linear systems, whereas a significant increase of the system effectiveness can be obtained for high amplitudes of vibrations and low frequencies of excitation signals,
- the experimental and simulation results indicate qualitative and quantitative compliance of dynamic properties of the analysed vibration reduction systems, which confirms that the adopted assumptions and simplifications in the modelling process are correctly formulated,
- the analysed properties of the exemplary vibration isolation system on the basis of a lumped parameter model are similar to the properties of a real object—a human body exposed to vibrations, which confirms the correctness of shaping the system characteristics on the basis of their simplified models.

9.2.2 Conclusions Regarding the Developed Methodology of Shaping the Vibration Isolation Properties

The most important methodological conclusions resulting from the present work are the following:

- by shaping vibration isolation properties, the increased efficiency of the vibration reduction systems can be obtained and the selection of dynamic characteristics of the system should be made appropriate for a priori assumed conditions of their work,
- shaping vibration isolation properties should be conducted taking into account each non-linearities of the physical phenomena occurring in vibration reduction systems, which enable to increase the reliability of the obtained results in respect to real systems,
- in order to obtain satisfactory effectiveness of reduction, the selection of system dynamic characteristics should be carried out especially for the specific spectral classes of excitation signals, which as a result enables to obtain the best performance of vibration reduction system for a given working condition,
- in order to access the possibility of subjective choice of the optimal performance characteristics by the operator, it is justified to determine various compromise solutions of the suspension system with respect to restrictions imposed on particular vibration isolation criteria,
- the application of the proposed optimisation method effectively facilitates a searching of the Pareto-optimal solutions which are required to define the dynamic characteristics of basic elements used in the vibration reduction system.

The strategy presented in this book constitutes an effective tool for supporting the process of designing vibration reduction systems whose application is provided to protect working machine operators. Its usefulness is confirmed by the results of simulation and experimental investigations, presented in this work, which are obtained for the developed solutions of suspension system and compared with the research results conducted on the basis of conventional system.

9.3 Directions for Further Research

The developed methodology of shaping vibration isolation properties supports the process of designing vibration isolation systems but does not exhaust complex issues connected with protection of working machine operators against harmful vibrations. The subject matter addressed in the book indicates the necessity to carry out further theoretical and experimental research in the scope of modelling, control and optimisation of the human isolation against vibration.

Further studies may aim to develop the methodology of shaping vibration isolation properties of systems with new excitation signals which generate oscillating motion. At the same time, the standardisation of their spectral characteristics is required, so as to reproduce real working conditions of machines and devices operated by humans. In the case of passive systems, it is recommended to propose new constructions of vibration isolators whose suitably shaped characteristics will enable the achievement of the best compromise between opposing vibration isolation criteria. In the case of semi-active and active systems, there exists an extensive area of investigating new control algorithms having complementary abilities of predictive, adaptive and robust control of vibration reduction systems. Apart from that, it would be worth considering the need to devise the next evaluation criteria regarding the effectiveness of vibration reduction systems. The criteria should take into account a higher number of various discomfort parameters describing harmful effect of vibration on the human body.

The methodology described in this book can also be used for shaping vibration isolation properties of vibration reduction systems which are applied in many other areas of modern technology. These areas include all applications of vibration isolation systems used to protect humans from harmful effects of mechanical vibrations, for example vehicle suspension systems, vibration control systems of buildings, vibration reduction systems in hand tools, etc.

MIX
Papier aus verantwortungsvollen Quellen
Paper from responsible sources
FSC® C105338

If you have any concerns about our products,
you can contact us on
ProductSafety@springernature.com

In case Publisher is established outside the EU,
the EU authorized representative is:
Springer Nature Customer Service Center GmbH
Europaplatz 3, 69115 Heidelberg, Germany

Printed by Libri Plureos GmbH
in Hamburg, Germany